FALKIRK DISTRICT LIBRARIES

BO'NESS

30124 01349208 9

CANCELLED

D1342130

BONESS

**Practical Electronics
Handbook**

Practical Electronics Handbook

Third Edition

IAN R. SINCLAIR

Newnes
An imprint of Butterworth-Heinemann Ltd
Linacre House, Jordan Hill, Oxford OX2 8DP

 PART OF REED INTERNATIONAL BOOKS

OXFORD   LONDON   BOSTON
MUNICH   NEW DELHI   SINGAPORE   SYDNEY
TOKYO   TORONTO   WELLINGTON

First published 1980
Reprinted 1982, 1983 (with revisions), 1987
Second edition 1988
Reprinted 1990
Third edition 1992

© Butterworth-Heinemann 1988, 1992

All rights reserved. No part of this publication
may be reproduced in any material form (including
photocopying or storing in any medium by electronic
means and whether or not transiently or incidentally
to some other use of this publication) without the
written permission of the copyright holder except in
accordance with the provisions of the Copyright,
Designs and Patents Act 1988 or under the terms of a
licence issued by the Copyright Licensing Agency Ltd,
90 Tottenham Court Road, London, England W1P 9HE.
Applications for the copyright holder's written permission
to reproduce any part of this publication should be addressed
to the publishers

British Library Cataloguing in Publication Data
Sinclair, Ian R.
  Practical Electronics Handbook. – 3Rev. ed
  I. Title
  621.381

ISBN 0 7506 0691 6

Library of Congress Cataloguing in Publication Data
Sinclair, Ian Robertson.
  Practical electronics handbook/Ian R. Sinclair. – 3rd ed.
    p. cm.
  Includes bibliographical references and index.
  ISBN 0 7506 0691 6
  1. Electronic circuits.   I. Title.
  TK7867.S5393   1992
  621.381–dc20   92–6743   CIP

Printed and bound in Great Britain by
Biddles Ltd, Guildford and King's Lynn

FALKIRK DISTRICT LIBRARIES

621.381   BN

Contents

Preface

Databooks often tend to be simply collections of information, with little or nothing in the way of explanation, and in many cases with so much information that the user has difficulty in selecting what he needs. This book has been designed to include within a reasonable space most of the information which is useful in electronics together with brief explanations which are intended to serve as reminders rather than instruction. The book is not, of course, intended as a form of beginner's guide to the whole of electronics, but the beginner will find here much of interest, as well as a compact reminder of electronic principles and circuits. The constructor of electronic circuits and the service engineer should both find the data in this book of considerable assistance, and the professional design engineer will also find that the items collected here are of frequent use, and would normally only be available in collected form in much larger volumes.

I hope therefore, that this book will become a useful companion to anyone with an interest in electronics, and that the information in the book will be as useful to the reader as it has been to me.

Ian R. Sinclair

Introduction

Mathematical Conventions

Quantities greater than 100 or less than 0.01 are usually expressed in the *standard form:* $A \times 10^n$ where A is a number less than 10, and n is a whole number. A positive value of n means that the number is greater than unity, a negative value of n means that the number is less than unity. To convert a number into standard form, shift the decimal point until a number between 1 and 10 is obtained and count the number of places which have been shifted, which will be the value of n. If the decimal point has had to be shifted to the left n is positive, if the decimal point had to be shifted to the right n is negative. For example:

1200 is 1.2×10^3, but 0.0012 is 1.2×10^{-3}

To convert numbers back from standard form, shift the decimal point n figures to the right if n is positive and to the left if n is negative. Example:

$5.6 \times 10^{-4} = 0.000\ 56$;　$6.8 \times 10^5 = 680\ 000$

Numbers in standard form can be entered into a scientific calculator by using the key marked Exp or EE (for details, see the manufacturer's instructions).

Numbers in standard form can be used for insertion in formulae, but component values are more conveniently written using the prefixes shown in *Table 0.1*. Prefixes are chosen so that values can be written without using small fractions or large numbers.

Throughout this book, equations have been given in as many forms as are normally needed, so that the reader should not have to transpose equations. For example, Ohm's law is given in all three forms, $V = RI$, $R = V/I$, and $I = V/R$. The units which must be used with formulae are also shown, and must be adhered to. For example, the equation: $X = 1/(2\pi fC)$ is used to find the reactance of a capacitor in ohms, using C in farads and f in hertz. If the equation is to be used with values given

Table 0.1

Prefix	Abbreviation	Power of Ten	Decimal multiplier
Giga	G	10^9	1 000 000 000
Mega	M	10^6	1 000 000
kilo	k	10^3	1 000
milli	m	10^{-3}	1/1 000
micro	μ	10^{-6}	1/1 000 000
nano	n	10^{-9}	1/1 000 000 000
pico	p	10^{-12}	1/1 000 000 000 000

Note that 1 000 pF = 1 nF; 1 000 nF = 1 μF and so on
Examples: 1 kΩ = 1 000 Ω (sometimes written as 1K0, see *Table 1.3*)
                    1 nF = 0.001 μF, 1 000 pF or 10^{-9} F
                    4.5 MHz = 4 500 kHz = 4.5 $\times$ 10^6 Hz

in μF and kHz, then values such as 0.1 μF and 15 kHz are entered as 0.1 $\times$ 10^{-6} and 15 $\times$ 10^3. Alternatively, the equation can be written as $X = 1/(2\pi fC)$ MΩ with f in kHz and C in nF.

In all equations, multiplication may be indicated by use of a dot (A.B) or by close printing ($2\pi fC$). Where brackets ( ) are used in an equation, the quantities within the brackets should be worked out first. For example:

$$2 (3 + 5) \text{ is } 2 \times 8 = 16 \quad \text{and} \quad 2 + (3 \times 5) \text{ is } 2 + 15 = 17$$

Transposing, or changing the subject of an equation, is simple provided the essential rule is remembered: an equation is not altered by carrying out identical operations on both sides.

Example: $Y = \dfrac{5aX + b}{C}$ is an equation

If this has to be transposed so that it reads as a formula to find X, then the procedure is to change both sides so that X is left isolated. The steps are as follows:

(a)   multiply both sides by C result: $CY = 5aX + b$

(b)   subtract b from both sides   result: $CY - b = 5aX$

(c)   divide both sides by $5a$ result: $\dfrac{CY - b}{5a} = X$

Now the equation reads $X = \dfrac{CY - b}{5a}$ which is the desired transposition.

Chapter 1

Passive Components

Resistors

Resistance, measured in ohms (Ω), is defined as the ratio of voltage (in volts) across a length of material to current (in amperes) through the material. When a graph is drawn of voltage across the material plotted

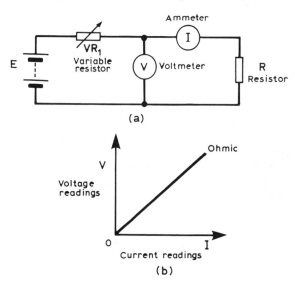

(a)

(b)

Figure 1.1. (a) A circuit for checking the behaviour of a resistor. (b) The shape of the graph of voltage plotted against current for an ohmic resistor, using the circuit in (a).

against current through the material, the value of resistance is represented by the *slope* of the graph. For a material which is kept at a constant temperature, a straight-line graph indicates that the material is *ohmic*, obeying Ohm's law (*Figure 1.1*). Non-ohmic behaviour is represented on such a graph by curved lines or lines which do not pass through the point, called the origin, which represents zero voltage and

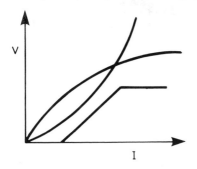

Figure 1.2. Three types of non-ohmic behaviour indicated by graph curves

zero current. Non-ohmic behaviour can be caused by temperature changes (light bulbs, thermistors), by voltage-generating effects (thermocouples), or by conductivity being affected by voltage (diodes), as in *Figure 1.2*.

Resistance values on components are either colour coded, as noted in *Table 1.2*, or have values printed on using the convention of BS 1852: 1970 (*Table 1.3*).

Resistivity

The resistance of any sample of material is determined by its dimensions and by the value of resistivity of the material. Wire drawn from a single reel will have a resistance value depending on the length cut; for example, a 3 m length will have three times the resistance of a 1 m length. When equal lengths of wire of the same material are compared, the resistance multiplied by the square of the diameter is the same for each. For example, if a given length of a sample wire has a resistance of 12 ohms and its diameter is 0.3 mm, then the length of wire of diameter 0.4 mm, of the same material, will have resistance R such that

$$R \times 0.4^2 = 12 \times 0.3^2 \text{ so that } R = \frac{12 \times 0.3^2}{0.4^2} = \frac{12 \times 0.09}{0.16} = 6.75 \text{ ohms}$$

Resistivity measures the effect which different materials contribute to the resistance of a wire. The resistivity of the material can be found by measuring the resistance R of a sample, multiplying by the area of cross-section, and then dividing by the length of the sample. As a formula, this is

$$\rho = \frac{R.A}{L}$$

where ρ (Greek rho) is resistivity, R is resistance, A is area of cross-section and L is length. When R is in ohms, A in square metres, and L

Table 1.2. RESISTOR COLOUR CODE

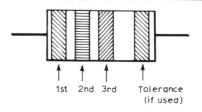

1st 2nd 3rd Tolerance
(if used)

Band Coding
First Band — first figure of resistance value (first *significant* figure)
Second Band — second figure of resistance value (second *significant* figure)
Third Band — number of zeros following second figure (multiplier)
Tolerance Band — percentage tolerance of value (5% or 10%). No tolerance band is used if the resistor has 20% tolerance.

Code Colours (also used for capacitor values)

Figure	Colour	
0	black	
1	brown	
2	red	
3	orange	
4	yellow	
5	green	
6	blue	
7	violet	
8	grey	
9	white	
0.01	silver	} used as multipliers (3rd band) only
0.1	gold	}

Tolerance	
10%	silver
5%	gold

Table 1.3. RESISTANCE VALUE CODING

Resistance values on components and in component lists are often coded according to BS 1852. In this scheme no decimal points are used, and a value in ohms is indicated by R, kilohms by K (not k), and megohms by M. The letter R, K or M is used in place of the decimal point, with a zero in the leading position if the value is less than 1 ohm.
Examples   1K5 = 1.5k or 1500 ohms; 2M2 = 2.2MΩ; 0R5 = 0.5 ohms

in metres, ρ is in ohm-metres. Since most wires are of circular cross-section, $A = \pi r^2$ (or $1/4\pi d^2$ where d is diameter).

Because the resistivities of materials are well known, this formula is much more useful in the form

$$R = \frac{\rho L}{A}$$

with ρ in ohm-metres, L in metres, A in metres^2, to find the resistance of a given length of wire of known area. This formula can be rewritten as

$$R = 1.27 \times 10^{-3} \frac{\rho L}{d^2}$$

with ρ in nano-ohm metres, as in *Table 1.4*, L in metres, and d (diameter) in millimetres. *Table 1.4* shows values of resistivities, in nano-ohm

Table 1.4. VALUES OF RESISTIVITY AND CONDUCTIVITY

The values of resistivity are in nano-ohm metres. The values of conductivity are in megasiemens per metres.

	Metal	*Resistivity*	*Conductivity*
	Aluminium	27.7	37
	Copper	17	58
	Gold	23	43
	Iron	105	9.5
Pure	Nickel	78	12.8
elements	Platinum	106	9.4
	Silver	16	62.5
	Tin	115	8.7
	Tungsten	55	18.2
	Zinc	62	16
	Carbon-steel (average)	180	5.6
	Brass	60	16.7
	Constantan	450	2.2
	Invar	100	10
	Manganin	430	2.3
Alloys	Nichrome	1105	0.9
	Nickel-silver	272	3.7
	Monel metal	473	2.1
	Kovar	483	2.0
	Phosphor-bronze	93	10.7
	18/8 stainless steel	897.6	1.11

metres, for various metals, both elements and alloys. The calculation of resistance by either formula follows the pattern of the example below.

Example A: Find the resistance of 6.5 m of wire, diameter 0.6 mm, if the resistivity value is 430 nano-ohm metres.

Using $R = \dfrac{\rho L}{A}$, $\rho = 430 \times 10^{-9}$, $L = 6.5$ m, $A = \pi \dfrac{(0.6 \times 10^{-3})^2}{4}$
(remembering that 1 mm = 10^{-3} m).

$$R = \frac{430 \times 10^{-9} \times 6.5}{2.82 \times 10^{-7}} = 9.88 \text{ ohms, about 10 ohms.}$$

Using $R = 1.27 \times 10^{-3} \dfrac{L}{d^2}$

$$R = \frac{1.27 \times 10^{-3} \times 430 \times 6.5}{0.36} = 9.86 \text{ ohms, nearly 10 ohms.}$$

To find the length of wire needed for a given resistance value, the formula is transposed to

$$L = \frac{RA}{\rho}, \text{ using } R \text{ in ohms, } A \text{ in metres}^2, \text{ and } \rho \text{ in ohm-metres}$$

to obtain L in metres. An alternative formula is

$$L = 785.4 \times \frac{Rd^2}{\rho} \text{ with } R \text{ in ohms, } d \text{ in millimetres, } \rho \text{ in nano-}$$
ohm metres.

To find the diameter of wire needed for a resistance R and length L metres, using ρ in nano-ohm metres, the formula is

$$d = 3.57 \times 10^{-2} \sqrt{\frac{\rho L}{R}} \text{ in millimetres}$$

For some purposes, conductivity is used in place of resistivity. The conductivity, symbol σ (Greek sigma), is defined as $\dfrac{1}{\text{resistivity}}$, so that $\rho = 1/\sigma$. The unit of conductivity is the siemens per metres, S/m. The resistivity formulae, using basic units, can be rearranged in terms of conductivity as follows

$$R = \frac{L}{\sigma A}, \qquad L = RA\sigma.$$

Conductivity values are also shown in *Table 1.4.*

Resistor Construction

The materials used for resistor construction are generally metal alloys, pure metal or metal oxide films, or carbon. Wirewound resistors use metal alloy wire wound onto ceramic formers. The winding must have a low inductance, so the wire is wound in the fashion shown in *Figure 1.3,* with each half of the winding wound in the opposite direction. Wire-wound resistors are used when very low values of resistance are needed, or when very precise values must be specified. Large resistance values, in the region of 20 kΩ to 100 kΩ need such fine-gauge wire that failure can occur due to corrosion in humid conditions, so that high-value wirewound resistors should not be used for marine or tropical applications unless the wire can be protected satisfactorily.

Figure 1.3. Non-inductive winding of a wirewound resistor. The two halves of the total length of wire are wound in opposite directions so that their magnetic fields oppose each other.

The majority of fixed resistors still use carbon composition, a mixture of graphite and clay whose resistivity value depends on the proportions of these materials used in the mixture. Because the resistivity value of such a mixture can be very high, greater resistance values can be obtained without the need for physically large components. Resistance value tolerances (see later) are high, however, because of the greater difficulty in controlling the resistivity of the mixture and the final dimensions of the carbon composition rod after heat treatment.

Metal film, carbon film and metal oxide film resistors are more recent types which are made by evaporating metals (in a vacuum) or tin oxide films (in air) onto ceramic rods. The resistance value is controlled (1) by controlling the thickness of the film; (2) by cutting a spiral pattern into the film after formation. These resistors are considerably cheaper to make than wirewound types, and can be made to closer tolerance values than carbon composition types.

Variable resistors and potentiometers can be made using all the methods that are employed for fixed resistors. The component is termed a potentiometer when connections are made to both ends as well as to the sliding connection; a variable resistor when only one end connection and the sliding connection is used. By convention, both are wired so that the quantity being controlled is *increased* by clockwise rotation of the shaft as viewed by the operator.

Any mass-production process aimed at producing one dimension will inevitably produce a range of values whose maximum tolerance can be specified. The tolerance is the maximum difference between any actual value and the target value, usually expressed as a percentage. For example a 10 kΩ 20% resistor may have a value of

$$10\ 000 + \left(\frac{20}{100} \times 10\ 000\right) = 12\ \text{k}\Omega \text{ or } 10\ 000 - \left(\frac{20}{100} \times 10\ 000\right) = 8\ \text{k}\Omega.$$

Tolerance series of preferred values (shown in *Table 1.5*), are ranges of target values chosen so that no component can be rejected. The mathematical basis of these preferred value figures is the sixth root of ten ($\sqrt[6]{10}$) for the E6 20% series (there are 6 steps of value between 1 and 6.8) and the twelfth root of 10 ($\sqrt[12]{10}$) for the E12 10% series. The figures produced by this series are rounded off. For example,

$$\sqrt[6]{10} = 1.46, \quad (\sqrt[6]{10})^2 = 2.15, \quad (\sqrt[6]{10})^3 = 3.16, \quad (\sqrt[6]{10})^4 = 4.64,$$
$$(\sqrt[6]{10})^5 = 6.8.$$

These figures are rounded to the familiar 1.5, 2.2, 3.3, 4.7, 6.8 used in the 20% series, and similar rounding is used for the 10% and 5% series.

A simpler view of the tolerance series is that, taking the 20% series as an example, 20% up on any value will overlap in value with 20% down on the next value. For example

4.7 + 20% = 5.64, and 6.8 − 20% = 5.44, allowing an overlap.

Table 1.5. PREFERRED VALUES TOLERANCE SERIES

E6 series (20%)	E12 series (10%)	E24 series (5%)
1.0	1.0	1.0
		1.1
	1.2	1.2
		1.3
1.5	1.5	1.5
		1.6
	1.8	1.8
		2.0
2.2	2.2	2.2
		2.4
	2.7	2.7
		3.0
3.3	3.3	3.3
		3.6
	3.9	3.9
		4.3
4.7	4.7	4.7
		5.1
	5.6	5.6
		6.2
6.8	6.8	6.8
		7.5
	8.2	8.2
		9.1

The numbers then repeat, but each multiplied by ten, up to 91Ω, then multiplied by 100 up to 910Ω and so on.

After manufacture resistors are graded with the 1%, 5% and 10% tolerance values removed. The remaining resistors are sold as 20% tolerance. Because of this, it is pointless to sort through a bag of 20% 6K8 resistors hoping to find one which will be of exactly 6K8 — such values will have been removed in the first sorting process by the manufacturer. Electronic circuits are designed to make use of wide tolerance components as far as possible. When close tolerance components are specified, it will be for a good reason, and 20% tolerance components cannot be substituted for 10% or 5% types.

Resistor characteristics

Important characteristics of resistor types include resistance range, usable temperature range, stability, noise level and temperature coefficient. Wirewound resistors are available in values ranging from fractions of an ohm (usually $0.22\,\Omega$) up to about $10\,k\Omega$; carbon compositions from about $2.2\,\Omega$ to $10\,M\Omega$, with film resistors available in ranges which are typically $1\,\Omega$ to $1\,M\Omega$. Typical usable temperature ranges are $-40°$ to $+105°C$ for composition, $-55°C$ to $+150°C$ for metal oxide. Wirewound resistors can be obtained which will operate at higher temperatures (up to $300°C$) depending on construction and value of resistance. The stability of value means the maximum change of value which can occur during shelf life, on soldering, or in adverse conditions, particularly operation at high temperature in damp conditions. Composition resistors have poorest stability, with typical shelf life change of 5%, soldering change of 2% and 'damp heat' change of 15% in addition to normal tolerance. Metal oxide resistors can have shelf life changes of 0.1%, soldering change of 0.15%, 'damp heat' changes of 1%, typically. The noise level of a resistor is specified in terms of microvolts (μV) of noise signal generated for each volt of d.c. across the resistor, and range from $0.1\,\mu V/V$ for metal oxide to a minimum of $2\,\mu V/V$ for composition (increasing with resistance value for composition resistors). The formula generally used for noise level of carbon composition resistors is $2 + \log_{10}\dfrac{R}{1\,000}\,\mu V/V$. For example, a $680\,k\Omega$ resistor would have a noise level of $2 + \log_{10}\dfrac{680\,000}{1\,000} = 2 + \log_{10}680\,\mu V/V = 4.8\,\mu V/V$.

The temperature coefficient of resistance measures the change of resistance value as temperature changes. The basic formula is

$$R_\theta = R_o\,(1 + a\,\theta)$$

where R_θ = resistance at temperature $\theta°C$, R_o = resistance at $0°C$, a = temperature coefficient, θ = temperature in $°C$.

The value of temperature coefficient is usually quoted in parts per million per $^\circ$C (ppm/$^\circ$C), and this has to be converted to a fraction (by dividing by one million) to use in the above formula.

Example: What is the value of a 6.8 kΩ resistor at 95°C, if the temperature coefficient is +1 200 ppm/$^\circ$C?

Solution: Converting +1 200 ppm/$^\circ$C to standard form,

$$= \frac{+1\ 200}{1\ 000\ 000} = 1.2 \times 10^{-3} \qquad (0.0012)$$

Using the formula: $R_0 = 6.8\ (1 + 1.2 \times 10^{-3} \times 95)$
   (The multiplication of $1.2 \times 10^{-3} \times 95$ must be carried out before adding 1).

$= 6.8\ (1 + 0.114)$

$= 7.57$ kΩ

Note that if the resistance at some other temperature ϕ is given (as distinct from the resistance at 0°C) then the formula becomes

$$R_0 = R\phi \left(\frac{1 + \alpha\theta}{1 + \alpha\phi} \right).$$ The 1s cannot be cancelled.

For example: if a resistor has a value of 10 Ω at 20°C, its resistance at 80°C can be found. If the value of α is 1.5×10^{-3}, then

$$R_0 = 10 \times \frac{1 + 80 \times 1.5 \times 10^{-3}}{1 + 20 \times 1.5 \times 10^{-3}} = 10 \times \frac{1 + .12}{1 + .03} = 10 \times \frac{1.12}{1.03} = 10.87\Omega$$

Temperature coefficients may be positive, meaning that the resistance will increase as the temperature rises, or negative meaning that the resistance will decrease as the temperature rises. Carbon composition resistors have temperature coefficients which vary from +1200 ppm/$^\circ$C, metal oxide types have the lowest temperature coefficient values of ±250 ppm/$^\circ$C.

Dissipation and temperature rise

The dissipation rating, measured in watts (W), for a resistor indicates how much power can be converted to heat without damage to the resistor. The rating is closely linked to the physical size of the resistor, so that ¼ W resistors are much smaller than 1 W resistors of the same resistance value. These ratings assume 'normal' surrounding (ambient) temperatures, and for high temperature use, derating must be applied according to the manufacturer's specification. For example, a ½ W component may have to be used in place of a ¼ W component when the ambient temperature is 70°C.

Figure 1.4 shows the graph of temperature rise plotted against dissipated power for average ½ W and 1 W composition resistors. Note that these figures are of temperature rise *above* the ambient level. If such a temperature rise takes the resistor temperature above the maximum temperature permitted in its type, a higher wattage resistor must be used.

The dissipation in watts is given by $W = VI$ with V the voltage across a conductor in volts and I the current through the conductor in amps.

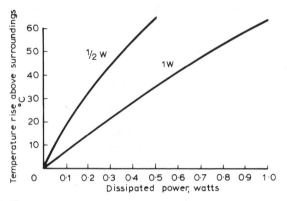

Figure 1.4. Temperature rise and power dissipation for typical carbon resistors. The temperature scale is in °C above surrounding (ambient) temperature. For example, in a room at 20°C, a ½W resistor dissipating 0.1 W will be at a temperature of 40°C

When current is measured in mA, then VI gives power dissipation in milliwatts. This expression for dissipated power can be combined with Ohm's law when the resistance of R of the conductor is constant, giving

$$W = V^2/R \text{ or } I^2R$$

The result will be in watts for V in volts and R in ohms or I in amps and R in ohms. When R is in kΩ, V^2/R gives W in milliwatts; when I is in mA and R in kΩ, W is also in milliwatts.

Note that *power* is energy transformed (from one form to another) per second. The unit of energy is the joule; the number of joules of energy dissipated is found by multiplying the power in watts by the time in seconds for which the power has been dissipated.

Variables and laws

The *law* of a variable resistor or potentiometer must be specified in addition to the quantities specified for any fixed resistor. The potentiometer law describes the way in which resistance between the slider and one contact varies as the slider is rotated; the law is illustrated by plotting a graph of resistance against shaft rotation angle (*Figure 1.5*). A linear law potentiometer (*Figure 1.5a*) produces a straight line graph, hence the name linear. Logarithmic (log) law potentiometers are

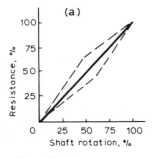 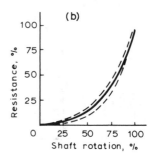

Figure 1.5. Potentiometer laws: (a) linear, (b) logarithmic. In the USA the word 'taper' is used in place of 'law', and 'audio' in place of 'log'. Broken lines show tolerance limits

extensively used as volume controls and have the graph shape shown in *Figure 5.1b*. Less common laws are anti-log and B-law; specialised potentiometers with sine or cosine laws are available.

Resistors in circuit

Resistors in circuit obey Ohm's law and Kirchhoff's laws. Ohm's law is written in its three forms as

$$V = RI; \qquad R = V/I; \qquad I = V/R$$

where V is voltage, R is resistance, I is current.

The units of these quantities are as shown in *Table 1.6*. These equations can be applied even to materials *which do not obey Ohm's law*, if the value of R is known. Materials which do not obey Ohm's law do not have a *constant* value of resistance, but the relationships given above hold good. The usefulness of the equations is greatest when the resistance values of resistors are constant.

Table 1.6. OHM'S LAW AND UNITS

Forms of the law: $V = RI$; $R = V/I$; $I = V/R$

Units of V	Units of R	Units of I
Volts, V	Ohms, Ω	Amps, A
Volts, V	Kilohms, kΩ	Milliamps, mA
Volts, V	Megohms, MΩ	Microamps, μA
Kilovolts, kV	Kilohms, kΩ	Amps, A
Kilovolts, kV	Megohms, MΩ	Milliamps, mA
Millivolts, mV	Ohms, Ω	Milliamps, mA
Millivolts, mV	Kilohms, kΩ	Microamps, μA

Kirchhoff's laws relate to the conservation of voltage and current. In a circuit, the voltage across each series component can be added to find the total voltage. Similarly the total current entering a junction must equal the total current leaving the junction. These laws are illustrated in *Figure 1.6.*

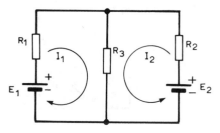

Current law: Current in $R_3 = I_1 + I_2$

Voltage law : $E_1 = R_1 I_1 + R_3(I_1 + I_2)$

$E_2 = R_2 I_2 + R_3(I_1 + I_2)$

Figure 1.6. Kirchhoff's laws. The current law states that the total current leaving a circuit junction equals the total current into the junction — no current is 'lost'. The voltage law states that the driving voltage (or e.m.f.) in a circuit equals the sum of voltage drops (IR) around the circuit

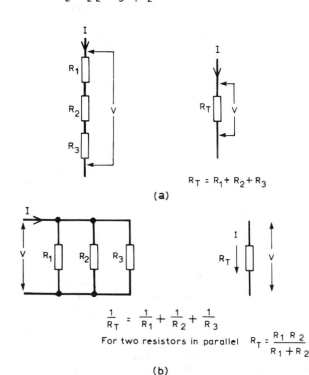

$R_T = R_1 + R_2 + R_3$

(a)

$$\frac{1}{R_T} = \frac{1}{R_1} + \frac{1}{R_2} + \frac{1}{R_3}$$

For two resistors in parallel $R_T = \dfrac{R_1 R_2}{R_1 + R_2}$

(b)

Figure 1.7. Resistors in series (a) and in parallel (b)

Figure 1.7 shows the rules for finding the total resistance of resistors in series or in parallel. When a combination of series and parallel connections is used, the total resistance of each series or parallel group must be found first before finding the grand total.

The superposition theorem is of great value in finding the voltages and currents in a circuit with two or more sources of voltage. *Figure 1.8* shows an example of the theorem in use. One supply is selected and the circuit redrawn to show the other supply or supplies short circuited. The voltage and current caused by the first supply can then be calculated, using Ohm's law and the rules for combining parallel resistors. Each supply is treated in this way in turn, and finally the currents and voltages caused by each supply are added.

SUPERPOSITION PRINCIPLE

In any linear network, the voltage at any point is the sum of the voltages caused by each generator in the circuit. To find the voltage caused by a generator, replace all other generators in the circuit by their internal resistances, and use Ohm's law.
A linear network means an arrangement of resistors and generators, with the resistors obeying Ohm's law and the generators having a constant voltage output and constant internal resistances.

Example: In the network shown, find the voltage across the 2.2kΩ resistor.

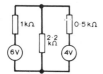

In this network, there are two generators and three resistors. The generators might be batteries, oscillators, or other signal sources.

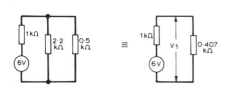

To find the voltage caused by the 6V generator, replace the 4V generator by its internal resistance of 0.5kΩ. Using Ohm's law, and the potential divider equation, V_1 = 1.736 V

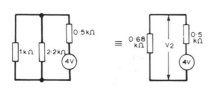

To find the voltage caused by the 4V generator, the 6V generator is replaced by its 1kΩ internal resistance. In this case, V_2 = 2.315 V

Now the total voltage in the original circuit across the 2.2kΩ resistor is simply the sum of these, 4.051 V.

Figure 1.8. Using the superposition theorem. This is a simple method of finding the voltage across a resistor in a circuit where more than one source is present.

16

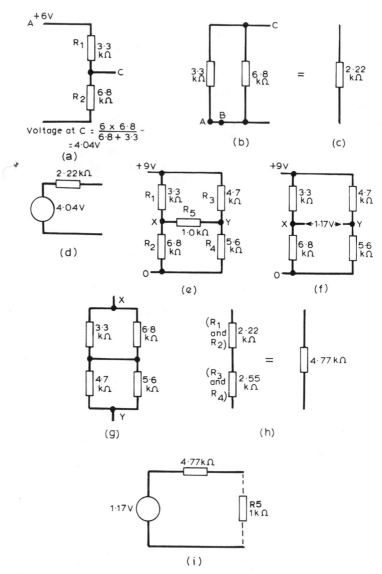

Figure 1.9. Using Thévenin's theorem. The potential divider (a) has an output voltage, with no load, of 4.04 V. It is equivalent to a 4.04 V source whose internal resistance is found by imagining the voltage supply short-circuited (b and c), so that the equivalent is as shown in (d). This makes it easy to find the output voltage when a current is being drawn. Similarly the bridge circuit (e) will have an open-circuit voltage, with R5 removed, of 1.17 V across X and Y (f), and the internal resistance between these points is found by imagining the supply short circuited (g). The combination of resistors in (g) is resolved (h) to give the single equivalent (i)

Thévenin's theorem is, after Ohm's law, one of the most useful electrical circuit laws. The theorem states that any linear network (such as resistors and batteries) can be replaced by an equivalent circuit consisting of a voltage source with a resistance in series. The size of the equivalent voltage is found by taking the open circuit voltage between two points in the network, and the series resistance by calculating the resistance between the same two points in the network, assuming that the voltage source is short circuited. Examples of the use of Thévenin's theorem are shown in *Figure 1.9.*

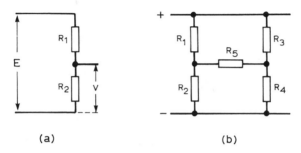

(a)   (b)

Figure 1.10. Potential divider (a) and bridge (b) circuits

Figure 1.10 shows two important networks, the potential divider and the bridge. When no current is taken from the potential divider, its output voltage is

$$V = \frac{R_2 E}{R_1 + R_2}$$

as shown, but when current is being drawn (as when a transistor is biased by this circuit) the equivalent circuit (using Thévenin's theorem) is more useful. The bridge circuit is said to be balanced when there is no voltage across R_5 (which is often a galvanometer or microammeter). In this condition,

$$\frac{R_1}{R_2} = \frac{R_3}{R_4}$$

If the bridge is *not* balanced, the equivalent circuit derived from Thévenin's theorem is, once again, more useful.

Thermistors

Thermistors are resistors made from materials which have large values of temperature coefficients. Both p.t.c. and n.t.c. types are produced for applications ranging from measurement to transient current suppression. Miniature thermistors in glass tubes are used for temperature measurement, using a bridge circuit (*Figure 1.11*), for timing, or for stabilising the amplitude of sine wave oscillators (see Chapter 3). Such thermistors are self heating if the current through them is allowed to exceed the limits laid down by the manufacturers, so that the current flowing in a bridge measuring circuit must be carefully limited. Larger thermistor types, with lower values of cold resistance (at 20°C) are

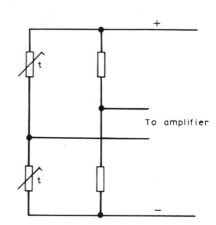

Figure 1.11. Thermistor bridge for temperature measurement

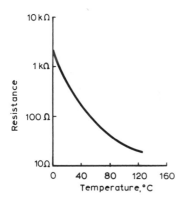

Figure 1.12. Graph of resistance plotted against temperature for typical thermistor

used for current regulation, such as circuits for degaussing colour TV tubes or controlling the surge current into light bulbs or valve heaters. The form of graphs of resistance plotted against temperature is that of *Figure 1.12* and the formula for finding the resistance at any temperature is shown in *Figure 1.13* with examples.

$$R_{\theta 1} = R_{\theta 2} \cdot e^{B\left(\frac{1}{\theta_1} - \frac{1}{\theta_2}\right)}$$

$R_{\theta 1}$ – resistance at temperature θ_1 (kelvins)

$R_{\theta 2}$ – resistance at temperature θ_2 (kelvins)

B – thermistor constant

Note: kelvin temperature = °C + 273

Calculator procedure:

Enter value of known temperature θ_1 then $\boxed{\frac{1}{x}}$, $\boxed{-}$, enter value of θ_2, $\boxed{\frac{1}{x}}$, $\boxed{=}$, $\boxed{x}$, enter value of B $\boxed{=}$ $\boxed{e^x}$ $\boxed{x}$ enter value of $R_{\theta 2}$

$\boxed{=}$ read answer

Example: A thermistor has a resistance of 47 kΩ at 20°C. what is its resistance at 100°C if its B value is 3900?

Equation is $R = 47 \times e^{3900\left(\frac{1}{373} - \frac{1}{293}\right)}$

$= 2 \cdot 7 \text{k}\Omega$

Figure 1.13. Finding thermistor resistance at any temperature, knowing the thermistor constant, B, and the resistance at a given temperature

Capacitors

Two conductors which are not connected and are separated by an insulator constitute a capacitor. When a cell is connected to such an arrangement, current flows *momentarily*, transferring charge (in the form of electrons) from one conducting plate (the + plate) to the other. When a quantity of charge Q has been transferred, the voltage across the plates equals the voltage V across the battery. For a given arrangement of conductors, the ratio $\frac{Q}{V}$ is a constant, and is called capacitance. The relationship can be written in three forms

$$Q = CV \qquad C = \frac{Q}{V} \qquad V = \frac{Q}{C}$$

The parallel-plate capacitor is the simplest practical arrangement, and its capacitance value is relatively easy to calculate. For a pair of parallel plates of equal area A, separation d, the capacitance is given by

$$C = \frac{\epsilon_r \epsilon_o A}{d}$$

The quantity ϵ_0 is called the permittivity of free space and has the constant value of 8.84×10^{-12} farads per metre.

Air has approximately this value of permittivity also, but other insulating materials have values of permittivity which are higher by the factor ϵ_r, which is different from each material. Values of this quantity, the relative permittivity (formerly called dielectric constant) are shown in *Table 1.7*.

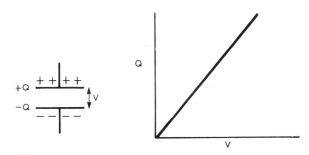

Figure 1.14. Basic principles of the capacitor. The relationship Q/V, shown in the graph, is defined as the capacitance, C

The formula can be recast, using units of cm^2 for area and mm for spacing, as

$$C = \frac{0.88 \times \epsilon_r \times A}{d} \text{ pF}$$

For small plate sizes, these units are more practical, but some allowance must be made for stray capacitance between any conductor and the metal surrounding it. Even a completely isolated conductor will have some capacitance.

Table 1.7.   VALUE OF ϵ_0 = 8.84 pF PER METRE

Material	Relative permittivity value
Aluminium oxide	8.8
Araldite resin	3.7
Bakelite	4.6
Barium titanate	600—1200 (varies with voltage)
Magnesium silicate	5.6
Nylon	3.1
Polystyrene	2.5
Polythene	2.3
PTFE	2.1
Porcelain	5.0
Quartz	3.8
Soda glass	6.5
Titanium dioxide	100

Example 1: Find the capacitance of two parallel plates 2 cm × 1.5 m, spaced by a 0.2 mm layer of material of relative permittivity 15.

Solution: Using $C = \dfrac{\epsilon_r \epsilon_o A}{d}$ ϵ_r = 15; ϵ_o = 8.84 × 10^{-12} F/m;

A = 0.02 × 1.5 m²; d = 0.2 × 10^{-3} m

$$C = \frac{15 \times 8.84 \times 10^{-12} \times 0.02 \times 1.5}{0.2 \times 10^{-3}} = 1.989 \times 10^{-8}\ \text{F}$$

about 2 × 10^{-8} F or 0.02 μF

Example 2: Find the capacitance of two parallel plates 2 cm × 1 cm spaced 0.1 mm apart by a material with relative permittivity 8.

Solution: Using $C = \dfrac{0.88 \times \epsilon_r \times A}{d}$ with A = 2 × 1 = 2 cm²,

d = 0.1 mm

$$C = \frac{0.88 \times 8 \times 2}{0.1}\ \text{pF} = 140.8\ \text{or}\ 141\ \text{pF}$$

Construction

Small value capacitors can be made using thin plates of insulating material metallised on each side to form the conductors. Thin plates can be stacked and interconnected (*Figure 1.15c*), to form larger

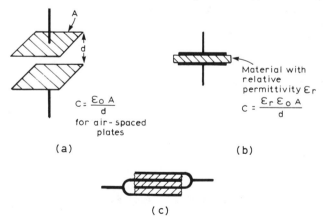

$$C = \frac{\epsilon_o A}{d}$$
for air-spaced plates

(a)

Material with relative permittivity ϵ_r

$$C = \frac{\epsilon_r \epsilon_o A}{d}$$

(b)

(c)

Figure 1.15. The parallel-plate capacitor. The amount of capacitance is determined by the plate area, plate separation, and the relative permittivity of the material between the plates

capacitance values up to 1000 pF or more. Silver mica types are used when high stability of value is important as in oscillators; and ceramic for less important uses (such as decoupling). Ceramic tubular capacitors make use of small tubes silvered inside and outside.

Rolled capacitors use as dielectric strips of paper, polyethylene, polyester, polycarbonate or other flexible insulators which are metallised (by vacuum coating) on each side and then rolled up (*Figure 1.16*),

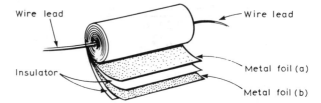

Figure 1.16. The rolled construction used for capacitors which make use of sheet dielectrics such as paper, polyester, polystyrene or polycarbonate

with another insulating strip to prevent the metallising on one side shorting against the metallising on the other side. Using this construction, quite large capacitance values can be achieved in small volume; up to a few μF is common.

Electrolytic capacitors are used when very large capacitance values are needed. One 'plate' is of aluminium in contact with an aluminium perborate solution in the form of a jelly or paste. The insulator is a film of aluminium oxide which forms on the positive 'plate' when a voltage, called the forming voltage, is applied during manufacture. Because the layer of oxide can be very thin (of only a few molecules thick) and the surface area of the aluminium can be made very large by roughening the surfaces, very large capacitance values can be achieved. The disadvantage of aluminium electrolytics include leakage current which is high compared to that of other capacitor types, polarisation (the + and − markings must be observed) and comparatively low voltage operation (less important when only transistor circuits are used). Incorrect polarisation can break down the oxide layer and if large currents can flow (as is the case in a power supply reservoir capacitor) the capacitor can explode, showering its surroundings with corrosive jelly.

Tantalum electrolytes can be used unpolarised (but not necessarily reverse polarised) and have much lower leakage currents that aluminium types, making them more suitable for some applications.

Capacitor characteristics

The same series of preferred values (20%) as is used for resistor values is also applied to values of capacitance (other than electrolytics) though older components will generally be marked with values such as 0.02 μF which are for all practical purposes equivalent to the preferred value 0.022 μF. Some manufacturers mark the values in pF only, using the rather confusing k to indicate thousands of pF (*Table 1.8*). Colour coded values are always in pF.

Electrolytic capacitors are always subject to very large tolerance values, of the order of −50% +100%, so that the actual capacitance value may range from half of the printed value to double that value. The insulation resistance between the plates is often so low that capacitance meters are unable to make accurate measurements. Capacitor values marked in circuit diagrams can use the BS 1852 method (6n8, 2μ2, etc) but are often marked in μF and pF. Quite commonly, fractions refer to μF and whole numbers to pF unless marked otherwise so that values of 0.02, 27, 1000, 0.05 mean 0.02 μF, 27 pF, 1000 pF and 0.05 μF respectively.

For all capacitors, the working voltage rating (abbreviated VW) must be carefully observed. Above this voltage, sparking between the

---

Table 1.8. CAPACITOR COLOUR CODING

Most capacitors are marked with values in μF or pF. The letter k is sometimes used in place of nF, i.e. 10 k = 10 nF = 0.01 μF. Colour coding is sometimes used:

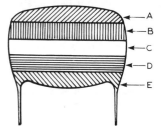

Bands A, B, C are used for coding values in pF in the same way as for resistors — remember that 1000 pF = 1 nF = 0.001 μF
Band D — Black = 20% White = 10%
Band E — Red — 250 V d.c. working
       Yellow — 400 V d.c. working

Colour code for small block capacitors (mainly polyester)

Tantalum electolytic capacitors are also sometimes colour coded, but with values in μF rather than pF

Band	1	2	3	4
Black	—	0	×1	10 V
Brown	1	1	×10	
Red	2	2	×100	
Orange	3	3	—	
Yellow	4	4	—	6.3 V
Green	5	5	—	16 V
Blue	6	6	—	20 V
Violet	7	7	—	
Grey	8	8	×0.01	25 V
White	9	9	×0.1	3V
Pink				35 V

conductors can break down the insulation, causing leakage current and eventual destruction of the capacitor. The maximum voltage that can be used is much lower at high temperatures than at low temperatures.

Values as low as 3 V may be found in high capacitance electrolytics; and values as high as 20 kV for ceramic capacitors intended for transmitters. The common voltage ranges used for capacitors in transistor circuits are shown in *Table 1.9.*

Table 1.9. CAPACITORS – COMMON WORKING VOLTAGES

10 V	16 V	20 V	25 V	35 V	40 V
63 V	100 V	160 V	250 V	400 V	1000 V

Changes of temperature and of applied d.c. voltage affect the value of capacitors because of changes in the dielectric. Both p.t.c. and n.t.c. types can be obtained, and the two are often mixed to ensure minimal capacitance change in, for example, oscillator circuits. Paper and polyester capacitors have, typically, positive temperature coefficients of around 200 ppm/$^{\circ}$C, but silver micas have much lower (positive) temperature coefficients. Aluminium electrolytics have large positive temperature coefficients, with a considerable increase in leakage current as temperature increases. In addition, electrolytics cannot be operated below about -20°C, as the electrolyte paste freezes. The normal working range for other types is -40°C to $+125^{\circ}$C, though derating may be needed at the higher temperature. Voltage ratings are generally for 70°C working temperature.

A few types of capacitors, notably 'High-K' ceramics, change value as the applied voltage is changed. Such capacitors are unsuitable for use in tuning circuits and should be used only for non-critical decoupling and coupling applications.

Variable capacitors can make use of variation of overlapping area or of spacing between plates. Air dielectric is used for the larger types (360 pF or 500 pF), but miniature variables make use of mica or plastic sheets between the plates. Compression trimmers are manufactured mainly in the smaller values, up to 50 pF. In use, the moving plates are always earthed, if possible, to avoid changes of capacitance (due to stray capacitance) when the control shaft is touched.

Energy and charge storage

The amount of charge stored by a capacitor is given by

$$Q = CV$$

When C is in μF and V in volts, Q is in microcoulombs (μC).

Example: How much charge is stored by a 0.1 μF capacitor charged to 50 V?

Solution: Using $Q = CV$ with C in μF, V in volts

$Q = 0.1 \times 50 = 5\ \mu$C.

When charged capacitors are connected to each other (but isolated from a power supply), the total charge is constant, equal to the sum of all charges on the capacitors. If the voltages are not equal, energy will be lost (as electromagnetic radiation) when the capacitors are connected. The amount of energy, in joules, stored by a charged capacitor is most conveniently given by $W = \frac{1}{2}CV^2$.

Other equivalent expressions are $\frac{1}{2}\dfrac{Q^2}{C}$ or $\frac{1}{2}QV$.

Example: How much energy is stored by a 5 μF capacitor charged to 150 V?

Solution: Using $\frac{1}{2}CV^2$, $C = 5 \times 10^{-6}$, $V = 150$

$W = \frac{1}{2} \times 5 \times 10^{-6} \times (150)^2$

$= 0.056$ J

This is used in connection with the use of capacitors to fire flash bulbs or in capacitor discharge car ignition systems.

In circuits, the laws concerning the series and parallel connections of capacitors are the *inverse* of those for resistors:

For capacitors in parallel: $C_{\text{total}} = C_1 + C_2 + C_3 + \ldots$

For capacitors in series: $\dfrac{1}{C_{\text{total}}} = \dfrac{1}{C_1}\ \dfrac{1}{C_2}\ \dfrac{1}{C_3}$

Time Constants

The charging and discharging of a capacitor is never instant. When a sudden step of voltage is applied to one plate of a capacitor, the other plate will step by the same amount. If a resistor is present, connecting the second plate to another voltage level, the capacitor will then charge to this other voltage level. The time needed for this change is about four time constants, as shown by *Figure 1.17*. The quantity, time constant T, is measured by $R \times C$ where R is the resistance of the charge/discharge resistor and C is the capacitance. For C in farads and R in ohms, T is in seconds. For the more practical units of μF and with resistance in kΩ, T is in milliseconds (ms) or for C in nF and resistance in kΩ, T is in microseconds (μs).

Example: In the circuit of *Figure 1.18*, how long does the voltage at the output take to die away?

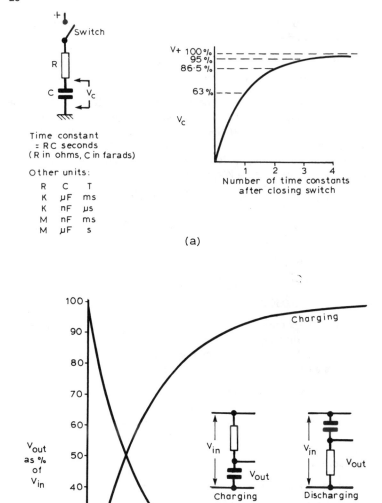

Figure 1.17. Capacitor charging and discharging. (a) Principles of charging, (b)
Universal charge/discharge curves

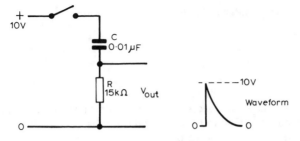

Figure 1.18. Time constant: differentiating circuit

Solution: With C = 0.01 μF = 10 nF and R = 15 kΩ, T = 150 μs.

Four time constants will be 4 $\times$ 150 μs = 600 μs, so that we can take it that the output voltage has reached zero after 600 μs.

Example: In the circuit of *Figure 1.19*, how long does the capacitor take to charge to 10 V?

Solution: With C = 0.22 μF = 220 nF and R = 6.8 kΩ, T = 6.8 $\times$ 220 = 1496 μs

Four time constants will be 4 $\times$ 1496 = 5984 μs or 5.98 ms, approx. 6 ms to charge.

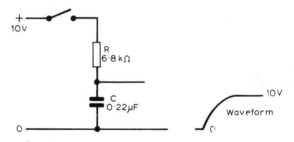

Figure 1.19. Time constant: integrating circuit

Note that a time of four time constants is taken for charging or discharging, because in practice charging or discharging is virtually complete by this time. The shape of the graph, however, indicates that charge is still being moved even several hundred time constants later.

Reactance

The reactance of a capacitor for a sine wave signal is given by

$$X_c = \frac{1}{2 \pi fC} \qquad (2 \pi = 6.28)$$

where C is capacitance in farads, f is frequency in hertz.

Reactance is measured in ohms, and is the ratio

$$\frac{\widetilde{V}}{\widetilde{I}}$$

where $\widetilde{V}$ is a.c. voltage across the capacitor and $\widetilde{I}$ is the a.c. current in the circuit containing the capacitor.

Unlike resistance, reactance is not a constant but varies inversely with frequency (*Figure 1.20*). In addition, the current sine wave is ¼ cycle (90°) ahead of the voltage sine wave across the capacitor plates. For phase and amplitude graphs of *CR* circuits, see later this chapter.

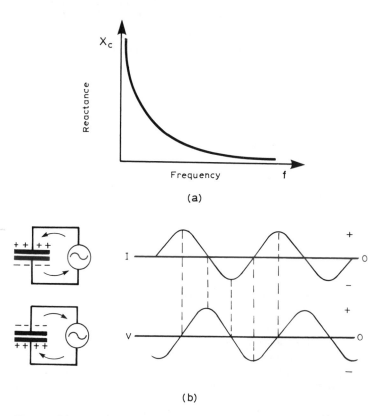

(a)

(b)

Figure 1.20. Capacitive reactance to a.c. signals. (a) Graph showing how capacitive reactance varies with frequency of signal. (b) Phase shift. As the capacitor charges and discharges, current flows alternately in each direction. The maximum current flow occurs when the capacitor is completely uncharged (zero voltage), and the maximum voltage occurs when the capacitor is completely charged (zero current). The graph of current is therefore ¼ cycle (90°) ahead of the graph of voltage

Inductors

An inductor is a component whose action depends on the magnetic field which exists around any conductor when a current flows through that conductor. When the strength of such a magnetic field (or magnetic flux) changes, a voltage is induced between the ends of the conductor. This voltage is termed an induced e.m.f., using the old term e.m.f. (electromotive force) to mean a voltage which has not been produced by current flowing through a resistor.

Faraday's Laws: Voltage induced depends on strength of magnet, speed of magnet (or coil) number of turns of coil, area of cross-section of coil

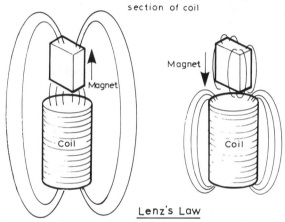

Lenz's Law

Magnet moving away from coil- magnetic field attracts magnet

Magnet moving towards coil coil- magnetic field repels magnet

The direction of induced e.m.f is such that it always opposes the change (movement in this case) which causes it

Figure 1.21. Faraday's and Lenz's laws. Faraday's laws relate to the size of the induced (generated) voltage in a coil to the strength, speed of the magnet and the size of the coil. Lenz's law is used to predict the direction of the voltage

If we confine our attention to static devices (coils and transformers rather than electric motors), then the change of magnetic field or flux can only be due to a change of current through one conductor. The induced e.m.f. is in such a direction that it opposes this change of current, and the faster the rate of change of current, the greater the opposing e.m.f. Because of its direction, the induced e.m.f. is called a *back e.m.f.* The laws governing these effects are Faraday's law and Lenz's law, summarised in *Figure 1.21.*

The size of the back e.m.f. can be calculated from the rate of change of current through the conductor and the details of construction of the conductor, straight wire or coil, number of turns of coil, use of a core etc. These constructional factors are lumped together as one quantity called inductance, symbol L. By definition,

$$E = L\frac{dI}{dt}$$

where E is the back e.m.f., L is inductance and $\frac{dI}{dt}$ is rate of change of current. The symbol 'd' in this context means 'change of' the quantity written after 'd'. If E is measured in volts and $\frac{dI}{dt}$ in amperes per second, then L is in henries (H).

Example: What back e.m.f. is developed when a current of 3A is reduced to zero in $\frac{1}{50}$ s (20ms) through a 0.5 H coil?

Solution: The amount of back e.m.f. is found from

$$E = L\frac{dI}{dt}$$

$$= 0.5 \times \frac{3}{20 \times 10^{-3}}$$

$$= 75 \text{ V}.$$

Note that this 75 V back e.m.f. will exist only for as long as the current is changing (20 ms), and may be *much* greater than the voltage drop across the coil when a steady current is flowing.

The rate of change of current is seldom uniform, so the back e.m.f. is usually a pulse waveform, whose maximum value must be found by measurement.

The existence of inductance in a circuit causes a reduction in the rate at which current can increase or decrease in the circuit. For a coil with inductance L and resistance R the time constant T is L/R seconds (L in henries, R in ohms). *Figure 1.22* shows how the current at a time t after switch-on varies in an inductive circuit — once again we take the time of four time constants to represent the end of the process.

The large e.m.f. which is generated when current is suddenly switched off in an inductive circuit can have destructive effects, causing sparking at contacts or breakdown of transistor junctions. *Figure 1.23* shows the commonly used methods of protecting switch contacts and transistor junctions from these switching transients.

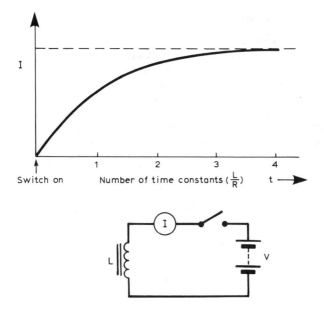

Figure 1.22. The growth of current in an inductive circuit

The changing magnetic field around one coil of wire will also affect windings nearby. The two windings are then said to have mutual inductance, symbol *M*. By definition

$$M = \frac{\text{back e.m.f. induced in second winding}}{\text{rate of change of current in first winding}}$$

using the same units as before, so that the unit of *M* is the henry.

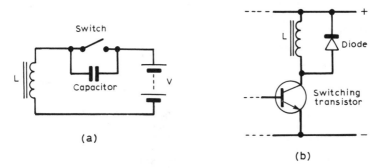

Figure 1.23. Protection against voltage surges in inductive circuits. (a) Using a capacitor across switch contacts, (b) using a diode across the inductor

Inductance calculations

Of all electronics calculations, those of inductance are the least precise. When an air-cored coil is used, the changing magnetic field does not affect all turns equally. Using a magnetic core makes the shape of the magnetic field more predictable, but makes its size less predictable. In addition, the permeability of the core changes considerably if d.c. flows in windings. Any equations for inductance are therefore very approximate, and should be used only as a starting point in the construction of an inductor. *Table 1.10* shows a formula for the number of

Table 1.10.   APPROXIMATE INDUCTANCE FOR AIR-CORED SINGLE-LAYER COIL (A SOLENOID)

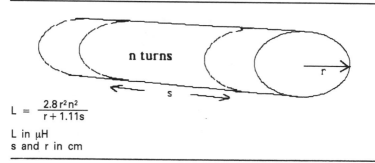

$$L = \frac{2.8\, r^2 n^2}{r + 1.11s}$$

L in μH
s and r in cm

turns of a single-layer close wound coil (solenoid) to achieve a given inductance. This approximate formula gives reasonable results for air-cored coils of values commonly used for radio circuit tuning. The addition of a core (generally of ferrite material) will cause an increase in inductance which could be by a factor as great as the relative permeability

Table 1.11.   RELATIVE PERMEABILITY VALUES

Relative permeability, $\mu_r = \dfrac{\text{Inductance of coil with core}}{\text{Inductance of coil without core}}$

Alternatively, inductance value with core = μ_r × inductance value without core.

Material	Relative permeability maximum value
Silicon-iron	7 000
Cobalt-iron	10 000
Permalloy 45	23 000
Permalloy 65	600 000
Mumetal	100 000
Supermalloy	1 000 000
Dustcores	10 to 100
Ferrites	100 to 2 000

(*Table 1.11*) of the ferrite. The multiplying effect is seldom as large as the value of relative permeability, because the ferrite does not normally enclose the coil. Manufacturers of ferrite cores which enclose coils

Example:

Inductance of a 120 turn coil is measured as 840µH

How many turns need to be removed to give 500µH ?

Since: $L \propto n^2$ (L- inductance, n- number of turns)

Then: $\dfrac{L_1}{L_2} = \dfrac{n_1^2}{n_2^2}$ and $\dfrac{840}{500} = \dfrac{120^2}{n_2^2}$

$\therefore \quad n_2^2 = \dfrac{120^2 \times 500}{840} = 8571$

$\therefore \quad n = \sqrt{8571} = 93$ approx

Figure 1.24. Adjusting inductors to different values

provide winding data appropriate for each type and size of core. *Figure 1.24* shows how inductors can be adjusted for a different inductance value, using the principle that inductance is proportional to the square of the number of turns.

Inductive reactance

The reactance of an inductor for a sine wave signal of frequency f hertz is $2\pi fL$, where L is inductance in henries. The reactance is the ratio $\tilde{V}/\tilde{I}$

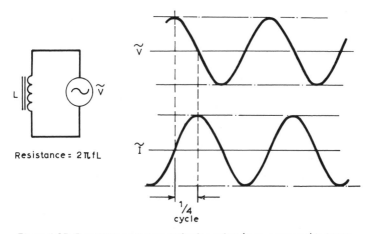

Resistance = $2\pi fL$

Figure 1.25. Reactance and phase shift of a perfect (zero resistance) inductor

($\widetilde{V}$ is signal voltage, $\widetilde{I}$ is signal current) and is measured in ohms. For a coil whose reactance is much greater than its resistance, the voltage sine wave is 90° (¼ cycle) ahead of the current sine wave (*Figure 1.25*).

Untuned transformers

An untuned transformer consists of two windings, primary and second-ary, neither or which is tuned by a capacitor, on a common core. For low frequency use, a massive core made from laminations (thin sections) of transformer steel alloy (such as silicon iron) must be used. Trans-formers which are used only for higher audio frequencies can make use of considerably smaller cores. At radio frequencies, the losses caused by transformer steels make such materials unacceptable, and ferrite materials are used as cores. For the highest frequencies, no core material is suitable and only self supporting air-cored coils (or pieces of straight wire or strip) can be used. In addition, high frequency currents flow mainly along the outer surfaces of conductors. This has two practical consequences — tubular conductors are as efficient as solid conductors (but use much less metal and can be water cooled) and silver plating can greatly decrease the effective resistance of a conductor.

For an untuned transformer with 100% coupling, the ratio of voltages $\widetilde{V}_s/\widetilde{V}_p$ is equal to the ratio of winding turns N_s/N_p, where $\widetilde{V}$ refers to a.c. voltage, N to number of turns and s, p to secondary, primary

Table 1.12. REACTIVE CIRCUIT RESPONSE

Circuit	$\dfrac{V_{out}}{V_{in}}$	Phase angle
	$\dfrac{1}{1 - f^2/f_0^2}$ $\approx -f_0^2/f^2$	0° when $f < f_0$ 180° when $f > f_0$
	$\dfrac{1}{1 - f_0^2/f^2}$ $\approx -f^2/f_0^2$	0° when $f > f_0$ 180° when $f < f_0$

Notes: f_0 is frequency of resonance $= \dfrac{0.16}{\sqrt{(LC)}}$

      f is frequency at which response is to be found
      $>$ greater than
      $<$ less than
      $\approx$ approximately equal to

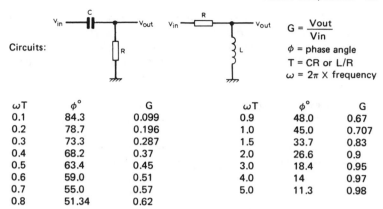

ωT	φ°	G		ωT	φ°	G
0.1	84.3	0.099		0.9	48.0	0.67
0.2	78.7	0.196		1.0	45.0	0.707
0.3	73.3	0.287		1.5	33.7	0.83
0.4	68.2	0.37		2.0	26.6	0.9
0.5	63.4	0.45		3.0	18.4	0.95
0.6	59.0	0.51		4.0	14	0.97
0.7	55.0	0.57		5.0	11.3	0.98
0.8	51.34	0.62				

Circuits:

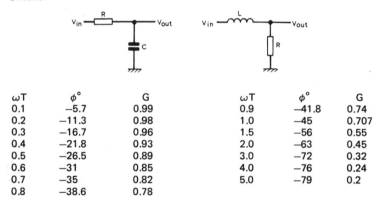

ωT	φ°	G		ωT	φ°	G
0.1	−5.7	0.99		0.9	−41.8	0.74
0.2	−11.3	0.98		1.0	−45	0.707
0.3	−16.7	0.96		1.5	−56	0.55
0.4	−21.8	0.93		2.0	−63	0.45
0.5	−26.5	0.89		3.0	−72	0.32
0.6	−31	0.85		4.0	−76	0.24
0.7	−35	0.82		5.0	−79	0.2
0.8	−38.6	0.78				

Figure 1.26. Amplitude/phase tables for LR and CR circuits. The quantity T is the time constant (CR or L/R), and ω is equal to 2πf

respectively. When an untuned transformer is used to transfer power between circuits of different impedance Z_p, Z_s, then the best match (maximum power transfer) condition is

$$\frac{N_s}{N_p} = \sqrt{\frac{Z_s}{Z_p}}$$

LCR circuits

The action of *CR* and *LR* circuits upon a sine wave signal is to change both the amplitude and the phase of the signal. Universal amplitude/ phase tables can be prepared, using the time constant *T* of the *CR* and

LR circuit and the frequency *f* of the sine wave. These tables are shown, with examples, in *Figure 1.26*.

When a reactance (*L* or *C*) is in circuit with a resistance *R*, the general formula for the total impedance (*Z*) are as shown in *Table 1.12*. Impedance is defined as $Z = \tilde{V}/\tilde{I}$, but the phase angle between *V* and *I* will not be 90°, and will be 0° only when resonance (see later) exists.

The combination of inductance and capacitance produces a tuned circuit which may be series (*Figure 1.27a*) or parallel (*Figure 1.27b*). Each type of tuned (or resonant) circuit has a frequency of resonance, symbol, f_o, at which the circuit behaves like a resistance so that there is no phase shift between voltage and current. At other frequencies, the circuit may behave like an inductor or like a capacitor. Below the frequency of resonance, the parallel circuit behaves like an inductor, the series circuit behaves like a capacitor. Above the frequency of resonance the parallel circuit behaves like a capacitor, the series circuit like an inductor. At resonance, the parallel circuit behaves like a large value resistor and the series circuit like a small value resistor.

The series resonant circuit can provide voltage amplification at the resonant frequency when the circuit of *Figure 1.28* is used. The amount of voltage amplification is given by $\dfrac{2\pi fL}{R}$ or $\dfrac{1}{2\pi fCR}$ at the frequency of

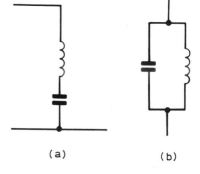

Figure 1.27. Tuned circuits (a) series, (b) parallel

(a)   (b)

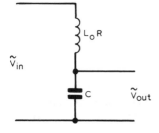

Figure 1.28. Voltage amplification of a tuned series circuit. The amplification is of the resonant frequency only, and can occur only if the signal source is of comparatively low impedance

resonance, and this quantity is often termed the circuit magnification factor, *Q*. There is no *power* amplification, as the voltage step up is achieved by increasing the current through the circuit, assuming that the signal input voltage is constant. *Table 1.12* shows typical phase and amplitude response formulae in universal form for the series resonant circuit. Note that the tuning capacitance may be a stray capacitance.

The parallel resonant circuit is used as a load which is a pure resistor (with no phase shift) only at the resonant frequency, f_o. The size of the equivalent resistance is called dynamic resistance and is calculated by the formula shown in *Figure 1.29*. The effect of adding resistors in

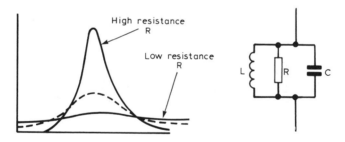

At resonance: $R_d = \dfrac{L}{CR}$

L = inductance in henries
C = capacitance in farads
R = resistance of coil in ohms

Figure 1.29. Dynamic resistance of a parallel resonant circuit

parallel with such a tuned circuit is shown in *Figure 1.30*, the dynamic resistance at resonance is reduced, but the resistance remains fairly high over a greater range of frequency. Such 'damping' is used to extend the bandwidth of tuned amplifiers. *Table 1.12* shows the amplitude and phase response of a parallel tuned circuit in general form. Once again, the tuning capacitance may be a stray capacitance.

Figure 1.30. The effect of damping resistance on the resonance curve

The impedance of a series circuit is given by the formula shown, with example, in *Table 1.12a*. Note that both amplitude (in ohms) and phase angle are given. The corresponding expression for a parallel circuit in which the only resistance is that of the coil (*R*) is also shown in *Table 1.13b*. When a damped parallel circuit is used, the resistance of the coil has generally a negligible effect compared to the damping resistor, and the formula of *Table 1.13c* applies.

Coupled tuned circuits

When two tuned circuits are placed so that their coils have some mutual inductance *M*, the circuits are said to be coupled. The size of the

Table 1.13. IMPEDANCE Z AND PHASE ANGLE ϕ

Circuit	Z	ϕ	
	$Z = \sqrt{\left[R^2 + \left(\omega L - \dfrac{1}{\omega C} \right)^2 \right]}$	$\phi = \tan^{-1} \left(\dfrac{\omega L - \dfrac{1}{\omega C}}{R} \right)$	(a)
	$Z = \sqrt{\left[\dfrac{R^2 + \omega^2 L^2}{(1 - \omega^2 LC)^2 + \omega^2 C^2 R^2} \right]}$	$\phi = \tan^{-1} \omega \left[\dfrac{L(1 - \omega^2 LC) - CR^2}{R} \right]$	(b)
	$Z = \dfrac{1}{\sqrt{\left[\left(\dfrac{1}{R} \right)^2 + \left(\omega C - \dfrac{1}{\omega L} \right)^2 \right]}}$	$\phi = \tan^{-1} R \left(\dfrac{1}{\omega L} - \omega C \right)$	(c)

Notes: ω = $2\pi \times$ frequency

$\tan^{-1}$ = angle whose tangent is equal to

mutual inductance is not simple to calculate; one approximate method is shown in *Table 1.14*. When the mutual inductance (*M*) between the coils is small compared to their self inductances (L_1, L_2) then the coupling is said to be loose, and the response curve shows a sharp peak. When the mutual inductance between the coils is large compared to their self inductances, the coupling is tight (or overcoupled) and the response curve shows twin peaks. For each set of coupled coils there is an optimum coupling at which the peak of the response curve is

Table 1.14. MUTUAL INDUCTANCE

1. From values of coil size *S* and *X* (note that these are lengths divided by coil diameter) find *K*.
2. Knowing inductance values, L_1, L_2, find *M*.
   $M = k\sqrt{(L_1 L_2)}$

S = spacing/diamater
X = coil winding length/diameter

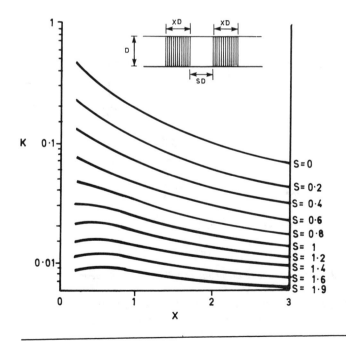

flattened and the sides steep. This type of response is an excellent compromise between selectivity and sensitivity.

The coefficient of coupling k is defined by

$$k = \frac{M}{\sqrt{(L_1 L_2)}}$$

(or $\frac{M}{L}$ if both coils have the same value of L) and critical coupling occurs when $k = \frac{1}{Q}$, assuming that both coils have the same Q factor — if they do not, then $Q = \sqrt{(Q_1 Q_2)}$. The size of the coefficient of coupling depends almost entirely on the spacing between the coils and no formulae are available to calculate this quantity directly.

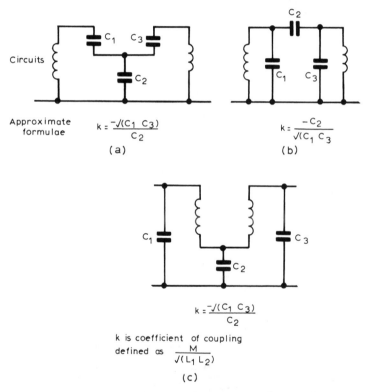

Circuits

Approximate formulae

$$k = \frac{-\sqrt{(C_1 C_3)}}{C_2}$$

(a)

$$k = \frac{-C_2}{\sqrt{(C_1 C_3)}}$$

(b)

$$k = \frac{-\sqrt{(C_1 C_3)}}{C_2}$$

k is coefficient of coupling defined as $\dfrac{M}{\sqrt{(L_1 L_2)}}$

(c)

Figure 1.31. Other methods of circuit coupling, and their design formulae

Other types of coupled circuits, with some design data, are shown in *Figure 1.31*. These make use of a common impedance or reactance for coupling and are not so commonly used.

Quartz crystals

Quartz crystals, cut into thin plates and with electrodes plated onto opposite flat faces, can be used as resonant circuits, with Q values ranging from 20 000 to 1 000 000 or more. The equivalent circuit of a crystal is shown in *Figure 1.32*. The crystal by itself acts as a series

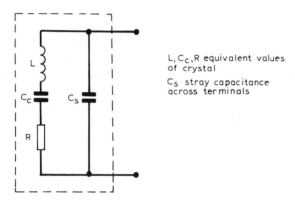

L, C_c, R equivalent values of crystal

C_s stray capacitance across terminals

Figure 1.32. Equivalent circuit of a quartz crystal

resonant circuit with very large inductance, small capacitance and fairly small resistance (a few thousand ohms). The stray capacitance across the crystal will also permit parallel resonance at a frequency slightly higher than that of the series resonance. *Figure 1.33* shows how the

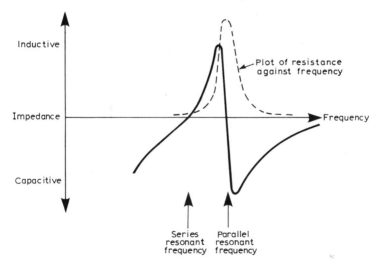

Figure 1.33. Variation of reactance and resistance of a crystal near its resonant frequencies

reactance and the resistance of a crystal vary as frequency is varied — the reactance is zero at each resonant frequency, the resistance is maximum at the parallel resonance frequency. Usually the parallel or the series frequency is specified when the crystal is manufactured.

Wave filters

Wave filter circuits are networks containing reactive components (L,C) which accept or reject frequencies above or below cut-off frequencies (which are calculated from the values of the filter components). Output amplitude and phase vary considerably as the signal frequency approaches a cut-off frequency, and the calculations involved are beyond the scope of this book. Much more predictable response can be obtained, for audio frequencies at least, by using active filters (see Chapter 3).

Measuring R, L, C

Resistance measurements can be made using the multimeter; the scale is non-linear but readings can be precise enough to indicate whether or not the value is within tolerance. Measurements of resistance, capacitance, self and mutual inductance can be carried out using bridge circuits (*Figure 1.34*) which rely on a 'null reading'. This means that potentiometers or switches are adjusted until a meter reading reaches a minimum,

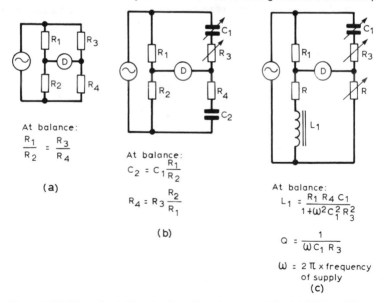

At balance:
$$\frac{R_1}{R_2} = \frac{R_3}{R_4}$$

(a)

At balance:
$$C_2 = C_1 \frac{R_1}{R_2}$$

$$R_4 = R_3 \frac{R_2}{R_1}$$

(b)

At balance:
$$L_1 = \frac{R_1 R_4 C_1}{1 + \omega^2 C_1^2 R_3^2}$$

$$Q = \frac{1}{\omega C_1 R_3}$$

$$\omega = 2\pi \times \text{frequency}$$
of supply
(c)

Figure 1.34. Measuring bridge circuits, with balance conditions. (a) Simple Wheatstone resistance bridge, (b) capacitance bridge, (c) inductance bridge

upon which the value of the quantity being measured can be read from the dials. Such bridge circuits use an audio frequency oscillator as a source of bridge voltage.

Direct-reading capacitance meters use a rather difference principle. Referring to *Figure 1.35*, the capacitor C is charged to a known voltage V, and then discharged through the meter, M. The amount of charge

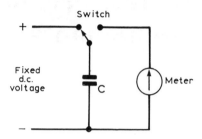

Figure 1.35. Principle of a direct-reading capacitance meter

passing through the meter on each discharge is CV, so that if the switch is actuated f times per second, the amount of charge flowing per second is fCV. This is the average current I, read by the meter, so that $I = fCV$ or $C = \dfrac{I}{fV}$. By a suitable choice of switching frequency, charging voltage and meter range, values of capacitance ranging from 10 pF to several μF can be measured, though erratic results are sometimes experienced with electrolytic capacitors. The switching is carried out by transistors, as the switching speeds needed are beyond the range of mechanical switches, even reed switches.

Chapter 2

Active Discrete Components

Diodes

Semiconductor diodes may use two basic types of construction, point contact or junction. Point contact diodes are used for small signal purposes where a low value of capacitance between the terminals is important — their main use nowadays is confined to r.f. demodulation. Junction diodes are obtainable with much greater ranges of voltage and current and are used for most other purposes. Apart from diodes intended for specialised purposes, such as light-emitting diodes, the fabrication materials are silicon or germanium, with germanium used almost exclusively for point contact demodulator diodes.

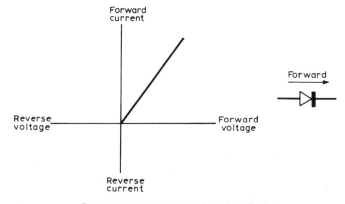

Figure 2.1. Characteristic of an ideal diode.

An ideal diode would conduct in one direction only, and the characteristic (the graph of current plotted against applied voltage) would look as in *Figure 2.1*. Practical diodes have a low forward resistance (not a *constant* value, however) and a high reverse resistance; and they conduct when the anode voltage is a few hundred millivolts higher than the cathode voltage.

The diode can be destroyed by excessive forward current, which causes high power dissipation at the junction or contact, or by using excessive reverse voltage, causing junction breakdown (see later), allowing it to conduct. Because reverse voltages are much higher than the voltage across a forward conducting diode, breakdown causes excessive current to flow so that once again the junction or contact is destroyed by excessive dissipation. For any diode therefore, the published ratings of peak forward current and peak reverse voltage should not be exceeded at any time, and should not be approached if reliable operation is to be achieved.

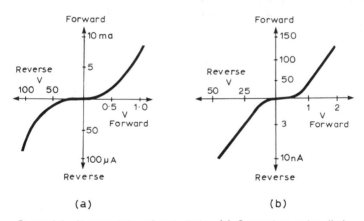

(a)   (b)

Figure 2.2. Characteristics of real diodes. (a) Germanium point diode,
(b) silicon junction diode. Note the different scales which have to be used
to allow the graphs to be fitted into a reasonable space

Characteristics for a typical point contact germanium diode and a typical small signal silicon junction diode are also shown in *Figure 2.2*. Comparing the two type of diode:

(a)   Germanium point contact diodes have lower reverse resistance values, conduct at a lower forward voltage (about 0.2 V) but have higher forward resistance because of their small junction area. They also have rather low peak values of forward current and reverse voltage.

(b)   Silicon junction diodes have very high values of reverse resistance, conduct at around 0.55 V forward voltage, can have fairly low forward resistance, and have fairly high peak values of forward current and reverse voltage.

The foward resistance of a diode is not a fixed quantity but is (very approximately) inversely proportional to current. Another approximation, useful for small currents, is that the forward voltage of a silicon diode increases by only 60 mV for a tenfold increase in current.

The effect of temperature change on a silicon diode is to change the forward voltage at any fixed value of current. A change of about 2.5 mV per $^\circ$C is a typical figure with the voltage reducing as the temperature is raised. The reverse (leakage) current is much more temperature dependent, and a useful rule of thumb is that leakage current doubles for each 10°C rise in temperature.

Zener diodes are used with reverse bias, making use of the breakdown which occurs across a junction when the reverse voltage causes a large electrostatic field across the junction. This breakdown limit occurs at low voltages (below 6 V) when the silicon is very strongly doped, and this type of breakdown is called zener breakdown. For such

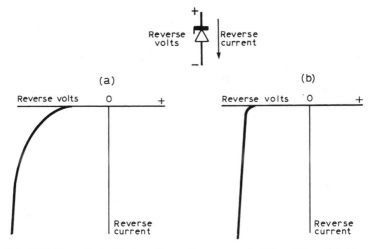

Figure 2.3. Zener diode. The true zener effect causes a 'soft' breakdown (a) at low voltages, the avalanche effect causes a sharper turnover (b)

a true zener diode, the reverse characteristic is as shown in *Figure 2.3a*. As the graph shows, the reverse current does not suddenly increase at the zener voltage, and the voltage across the diode is not truly stabilised unless the current is more than a few milliamps. This type of characteristic is termed a 'soft' characteristic. In addition to this, a true zener diode has a negative temperature coefficient — the voltage across the diode (at a constant current value) decreases as the junction temperature is increased.

Avalanche breakdown occurs in diodes with lower doping levels, at voltages above about 6 V. The name is derived from the avalanche action in which electrons are separated from holes by the electric field across the junction, and these electrons and holes then cause further electron-hole separation by collision. These diodes have 'hard' characteristics (*Figure 2.3b*) with very little current flowing below the avalanche voltage and large currents above the avalanche voltage. In addition, the temperature coefficient of voltage is positive, so that the voltage across the diode *increases* as the junction temperature is raised.

Both types of diodes are known, however, as zener diodes, and these with breakdown voltages between 4 V and 6 V combine both effects. At a breakdown voltage of around 5.6 V, the opposing temperature characteristics balance, so that the breakdown voltage of a 5.6 V diode is practically unaffected by temperature. The stabilisation of the diode is measured by its dynamic resistance, defined as the ratio

$$\frac{\text{Voltage change, } V}{\text{Current change, } I}\text{ , units of ohms,}$$

when V is the change of voltage across the diode caused by a change of current I when the diode is stabilising.

This ratio should be below 50 ohms, and reaches a minimum value of about 4 ohms for a diode with a breakdown voltage of about 8 V.

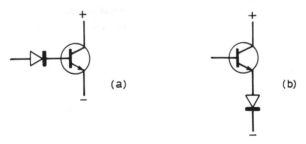

Figure 2.4. Protecting the base-emitter junction of a transistor against excessive reverse voltage, (a) diode in base circuit, (b) diode in emitter circuit. The base circuit is preferred, since the diode does not have to pass the emitter current of the transistor

Note that the base-emitter junction of many types of silicon transistors will break down by avalanche action at voltages ranging from 7 V to 20 V reverse bias, though this action does not necessarily cause collector current to flow. The base-emitter junction can be protected by a silicon diode (with a high breakdown voltage) wired in series (*Figure 2.4*).

Reference diodes are doped to an extent which makes the breakdown voltage practically constant despite changes in temperature. Voltages of 5 V to 6 V are used, and temperature changes ranging from $\pm$ 0.01% per degree to 0.000 5% per degree can be achieved. Reference diodes are used for very precise voltage stabilisation.

Varactor Diodes

All junction diodes have a measurable capacitance between anode and cathode when reverse biased, and this capacitance varies with voltage, being least when high reverse voltage are used. This variation is made use of for varactor diodes, in which the doping is arranged for the maximum possible capacitance variation. A typical variation is of 10 pF at 10 V bias to 35 pF at 1 V reverse bias. Varactor diodes are used for electronic tuning applications; a typical circuit is shown in *Figure 3.26* (Chapter 3).

LEDs

Light emitting diodes use compound semiconductors such as gallium arsenide or indium phosphide. When forward current passes, light is emitted from the junction. The colour of the light depends on the material used for the junction, and the brightness is approximately proportional to forward current. LEDs have higher forward voltages when conducting; 1.6 to 2.2 V as compared to the 0.5 to 0.8 V of a silicon diode. The maximum permitted reverse voltages are low, typically 3 V, so that a silicon diode must be connected across the LED as

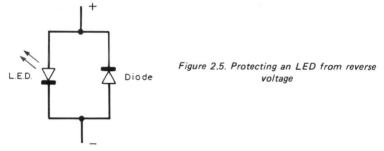

Figure 2.5. Protecting an LED from reverse voltage

shown in *Figure 2.5* if there is any likelihood of reverse voltage (or a.c.) being applied to the diode. A series resistor must always be used to limit the forward current unless pulsed operation is used.

For more specialised applications, microwave diodes of various types can be obtained which emit microwave radiation when forward biased and enclosed in a suitable resonant cavity. Tunnel diodes are diodes with an unstable portion of the characteristic (a reverse slope, indicating 'negative resistance'). When the tunnel diode is biased to the unstable region, oscillations are generated at whatever frequency is permitted by the components connected to the diode (*RC, LC,* cavity, etc.)

Diode circuits

Figure 2.6 shows some application circuits for diodes, with approximate design data where appropriate. Diode types should be selected with reference to the manufacturers data sheets, having decided on the basic reverse voltage and load current quantities required by the circuit.

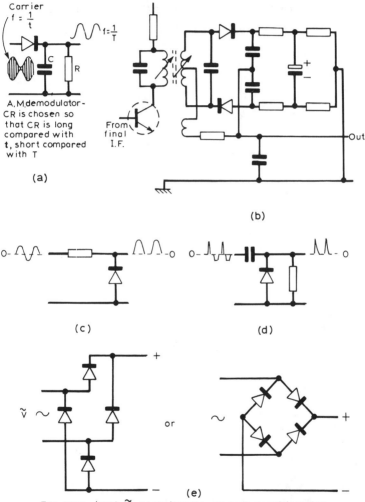

For r.m.s. input $\tilde{V}$, peak inverse on each rectifier diode equals 1·57V, ripple frequency is twice line frequency, average current per diode is equal to r.m.s. current

Figure 2.6. Some diode applications: (a) amplitude demodulation, (b) ratio detector for f.m. (c) signal clipping, (d) d.c. restoration, (e) bridge rectification

Transistors

Like signal diodes, transistors may be constructed using either silicon or germanium, but virtually all transistors now use silicon. The design data of this section refer to silicon transistors only.

The working principle of a transistor is that current flows between the collector and emitter terminals only when current is flowing between the base and emitter terminals. The ratio of these currents is called

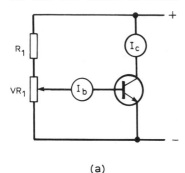

(a)

Figure 2.7. Forward current transfer ratio (a) measuring circuit, (b) graph. The slope of the graph (I_c/I_b) is equal to the forward current transfer ratio, h_{fe}

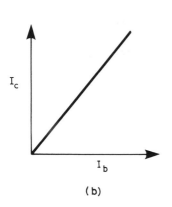

(b)

forward current transfer ratio, symbol h_{fe}. For the arrangement of *Figure 2.7*, the ratio is

$$h_{fe} = \frac{i_c}{i_b}$$

In databooks, a distinction is made between h_{FE}, for which I_c and I_b are d.c. quantities; and h_{fe}, for which i_c and i_b are a.c. quantities. The two quantities are however generally close enough in value to be interchangeable, and the symbol h_{fe} will be used here to indicate both values. The size of h_{fe} for any transistor can be measured in the circuit of *Figure 2.7a*; a simpler method, used in many transistor testers, is shown in *Figure 2.8*. Values vary from about 25 (power transistors operating at high currents) to over 1000 (some high-frequency amplifier types).

Base current will not flow unless the voltage between the base and the emitter is correct. The precise voltage at which current starts to

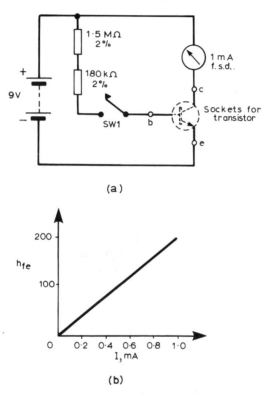

(a)

(b)

Figure 2.8. A simple transistor tester (a) and its calibration graph (b)

become detectable varies from one specimen (even of the same type number) of transistor to another, but for silicon transistors is generally about 0.5 V. The pnp type of transistor will require the emitter to be at a more positive voltage than the base, the npn type will require the base to be more positive than the emitter. When the transistor has the correct d.c. currents flowing (with no signal applied) it is said to be correctly biased. Amplification is carried out by adding a signal voltage to the steady voltage at the input of the transistor. The vast majority of transistor circuits use the base as the input terminal, though a few (common base amplifiers) use the emitter as an input terminal.

Bias for linear amplifiers

A linear amplifier produces at its output a waveform which is a perfect copy, but of greater amplitude, of the waveform at the base. The voltage gain of such an amplifier is defined as

$$G = \frac{V_{out}}{V_{in}}$$

where V indicates a.c. (signal) voltage measurements. If the output waveform is not a perfect copy of the input, then the amplifier exhibits distortion of one sort or another. One type of distortion is non-linear distortion, in which the shape of the waveform is changed by the 'copying' process. Such non-linear distortion is caused by the action of the transistor and can be minimised by careful choice of transistor type (see later) and by correct bias.

A transistor is correctly biased when the desired amount of gain can be obtained with minimum distortion. This is easiest to achieve when the (peak to peak) output signal from the amplifier is much smaller than the supply voltage. This may be achieved with the no-signal voltage at the collector of the transistor at almost any reasonable level, but to allow for unexpected signal overloads, the preferred collector voltage is half-way between supply positive voltage and the voltage of the emitter.

When the value of collector resistor has been chosen, bias is applied by passing current into the base so that the collector voltage drops to the desired values of $0.5\ V_{ss}$ where V_{ss} is supply voltage. For any bias system, the desired base current must be equal to

$$\frac{0.5\ V_{ss}}{R_L \times h_{fe}}\ \text{mA, with } V_{ss} \text{ in volts, } R_L \text{ in k}\Omega, h_{fe} \text{ as a ratio.}$$

Figure 2.9 shows three bias systems, with design data for obtaining a suitable bias voltage. The method of *Figure 2.9a* is the most difficult to use, as a different resistor value must be chosen for each transistor used. It may be necessary to use resistors in series or in parallel to achieve the correct value and the collector voltage will decrease noticeably as the temperature of the transistor increases. The method of *Figure 2.9b* is a considerable improvement over that of *Figure 2.9a*. The bias system may be designed around an 'average' transistor (with an average value of h_{fe} for that type) and can then be used unchanged for other transistors without too serious a change in the collector voltage. In addition the collector voltage changes much less as the temperature changes.

The bias system of *Figure 2.9c* is one which can be used for any transistor provided that the current flowing through the two base bias resistors R_1, R_2 is much greater than the base current drawn by the transistor. Unlike the other two systems, the design formula does not

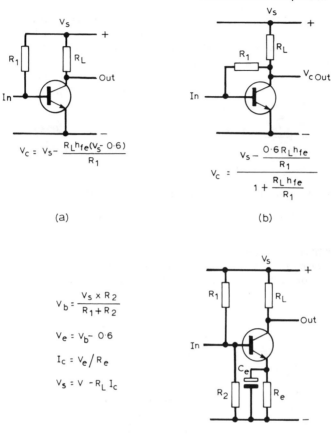

Figure 2.9. Transistor bias circuits: (a) simple system, usually unsatisfactory;
(b) using negative feedback of bias and signal; (c) potential divider method

require the h_{fe} value of the transistor to be known if the standing
current through the transistor is to be only a few milliamps. For power
transistors, the quantities that are needed are the V_{be} and I_{be} at the
bias current required. This system does not, however, stabilise the
collector voltage so effectively against changes caused by changes of
temperature.

Transistor parameters and linear amplifier gain

Transistor parameters are measurements which describe the action of
the transistor. The name parameter is used to distinguish these quantities

from constants. Transistor parameters are not generally constants, they vary from one transistor to another (even of the same type) and from one value of bias current to another. One such parameter, the common-emitter current gain, h_{fe}, has already been described.

Of the parameters for linear amplifiers, G_m is probably the most useful. G_m, called mutual conductance and measured in units of milli-siemens (mS) (equal to milliamps per volt) is defined by

$$G_m = \frac{\tilde{I}_c}{\tilde{V}_{be}}$$

where I_c = a.c. signal current, collector to emitter

where V_{be} = a.c. signal voltage, base to emitter.

The usefulness of G_m as a parameter arises from the fact that the voltage gain of a transistor amplifier for small signals is given by

$G = G_m R_L$ where R_L is the load resistance for signals (if G_m is measured in mS and R_L in kΩ, gain is correctly specified)

Note that R_L will generally be less than the resistance connected between the collector and the positive supply, because this value will be shunted by any other resistors connected through a capacitor to the collector

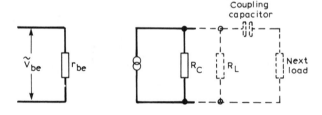

Current generator has infinite resistance and generates a signal current equal to $G_m \tilde{V}_{be}$

Figure 2.10. A useful equivalent circuit for the transistor. The signal voltage $\tilde{V}_{be}$ between the base and the emitter causes an output signal current $G_m \tilde{V}_{be}$. This current flows through the parallel resistor R_C (the transistor output resistance), R_L, the load resistor, and any other load resistors in the circuit

(*Figure 2.10*). The input resistance of the next transistor (if used) will also be in parallel with the collector resistor.

A graph of collector current plotted against base-emitter voltage is not a straight line, so that the ratio $\frac{i_c}{V_{be}}$ which is G_m, is not constant. A useful rule of thumb for small bias currents is that G_m = 40 $\times$ bias current in milliamps with G_m in mS (mA/V). The shape of the graph is always curved for low currents, but can vary in shape at higher currents. For a few transistors, the i_c, V_{be} graph has a noticeably straight portion, making these transistors particularly suitable for linear amplification applications. It is this straightness of the G_m characteristic which makes some types of power transistor much more desirable (and high priced) for audio output stages than others.

To take advantage of these linear characteristics, of course, the bias must be arranged so that the working point is at the centre of the linear region with no signal input. 'Working point' in this context means the combination of collector volts and base volts which represents a point on the characteristic.

Two other useful parameters for silicon transistors used in the common-emitter circuit are the input and output resistance values. The input resistance is defined as

$$\frac{\text{signal voltage, base input}}{\text{signal current into base}}$$

with no signal at the collector, symbol h_{ie}.

Then output resistance is, using a similar definition

$$\frac{\text{signal voltage at collector}}{\text{signal current at collector}}$$

with no signal at the base, symbol h_{oe}.

The output resistance h_{oe} has about the same range of values, from 10 kΩ to 50 kΩ for a surprisingly large number of transistors, irrespective of operating conditions, provided these are on the flat part of the V_{ce}/I_{ce} characteristic (*Figure 2.11*). An average value of 30 kΩ can usually be taken.

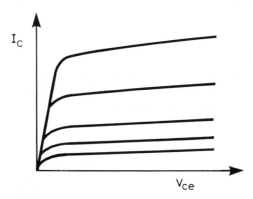

Figure 2.11. The I_c/V_{ce} characteristic. The flat portion is the operating part. The small amount of slope indicates that the output resistance, R_c, is high, usually 40 kΩ or more

The input resistance is not a constant, because the input of a transistor is a diode, the base-emitter junction. The value of input resistance h_{ie} is related to steady bias current, and to the other parameters

$$h_{ie} = \frac{h_{fe}}{G_m} \quad \text{and since } G_m = 40I_c,$$

$$= \frac{h_{fe}}{40I_c} \quad \text{where } I_c \text{ is the steady, no-signal, bias collector current.}$$

For example: if a transistor has an h_{fe} value of 120, and is used at a current of 1 mA, then the input resistance (in kΩ) is

$$h_{ie} = \frac{h_{fe}}{40I_c} = \frac{120}{40 \times 1} = 3k\Omega.$$

Noise

Any working transistor generates electrical noise, and the greater the current flowing through the transistor the greater the noise. For bipolar transistors, the optimum collector current for low noise operation is given approximately (in milliamps) by

$$I_c = \frac{28 \sqrt{h_{fe}}}{R_g} \quad \text{where } R_g \text{ is the signal source resistance in ohms.}$$

Low-noise operation is most important for the first stages of audio preamplifiers and for r.f. tuner and early i.f. stages. The noise generated by large value resistors is also significant, so that the resistors used for small signal input stages should be fairly low value, high stability film types, with small currents flowing. Variable resistors must never be used in a low-noise signal stage.

Voltage gain

The voltage gain of a simple single stage voltage amplifier can be found from a simple rule of thumb. If V_{bias} is the steady d.c. voltage *across the collector load resistor*, then the voltage gain is

$$G = 40 \times V_{bias}$$

For a single stage amplifier, the signal is attenuated both at the input and at the output by the resistance of devices connected to the transistor (microphones, tape heads, other amplifying stages). If the resistance of the signal source is R_s and the resistance of the next stage is R_{load}, then the measured gain will be

$$G \times \frac{h_{ie}}{R_s + h_{ie}} \times \frac{R_{load}}{R_{load} + R_c}$$

where R_c is the collector output resistance as shown in *Figure 2.12*, and G is the value of gain given by $40V_{bias}$. This method gives gain values accurately enough for most practical purposes. When precise values of gain are needed, negative feedback circuits (see below) must be used.

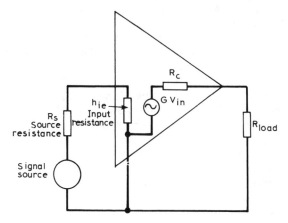

Figure 2.12. The voltage signal equivalent. The voltage gain, G, obtained by the transistor is reduced by the potential divider networks at the input and at the output

For a multi-stage amplifier, the gains of individual stages are multiplied together, and the attenuations caused by the potential dividing actions of R_s and R_{load} are also multiplied together.

FETs

Field effect transistors (FETs) are constructed with no junctions in the main current path between the electrodes, which are called drain and source respectively. The path between these contacts, called the channel, may be p-type or n-type silicon, so that both p-channel and n-channel FETs are found. Control of the current flowing in the

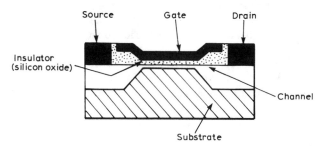

Figure 2.13. MOSFET structure

channel is achieved by varying the voltage at the third electrode, the gate. In junction FETs, the gate is a contact to a junction formed on the channel and reverse biased in most circuit applications. MOSFETs use a capacitor structure (*Figure 2.13*), so that the gate is completely insulated from the channel. No bias is needed, since the gate is insulated, but care has to be taken to avoid gate breakdown caused by excess voltage. Even electrostatic voltages, generated by contact, can cause damage, so that the gate electrode should be shorted to the source until the MOSFET is wired into circuit. In any circuit application, there must be a resistor connected from gate to source.

By altering the geometrical shape of the FET, power output FETs can be constructed. These are usually known as VFETs, the 'V' meaning 'vertical' and describing the construction, which is arranged so that the drain can be large and easily put into contact with a heat sink. Matched complementary pairs of these VFETs have been used to a considerable extent in hi-fi amplifiers.

The input resistance of either type of FET is very high, and low noise levels can be achieved, even with source resistances as high as 1MΩ.

Negative feedback

Feedback means using a fraction of the output voltage of the amplifier as an input. When the signals at input and output are oppositely phased (mirror-image waveform), then the feedback signal is said to be negative. Negative feedback signals subtract from the input signals to the amplifier, so reducing the overall gain of the amplifier. The effect on gain is as follows:

Let G = Gain of amplifier with no feedback, known as the open loop gain

n = feedback fraction (or loop gain), so that V_{out}/n is fed back.

Then the gain of the amplifier when negative feedback is applied is

$$G/(1 + \frac{G}{n}) = \text{the closed loop gain.}$$

For example, if open loop gain, $G = 100$ and $n = 20$ (so that 1/20 of the output voltage is fed back) then the closed loop gain is $\dfrac{100}{1 + \dfrac{100}{20}} = \dfrac{100}{6} =$

16.7. A very useful approximation is that if the open loop gain G is very much greater than the loop gain, then the closed loop gain (with the negative feedback connected) is simply equal to n. This is because G/n is large, so that in the expression above it is much greater than unity and the expression is approximately $G/(G/n) = n$.

Negative feedback, in addition to reducing gain, reduces noise signals which originate in the components of the amplifier, and will also

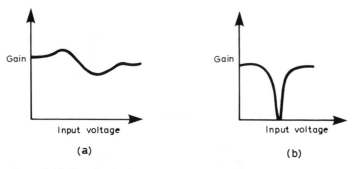

Figure 2.14. *Negative feedback can correct non-linear distortion provided that the gain (before feedback) remains reasonably high (a) over the full range of input voltages. If the gain is zero for any input (b) or is unusually low, then feedback cannot correct the distortion. Cross-over distortion is an example of a fault which causes zero gain.*

reduce distortion provided that the distortion does not cause loss of open-loop gain (*Figure 2.14*). Input and output resistances are also affected in the following way. If the feedback signal shunts the input (*Figure 2.15*) (applied to the same terminal), then input resistance is reduced, often to such an extent that the input terminal is practically at earth potential for signals (a *virtual* earth). If the feedback signal is in series with the input signal (*Figure 2.16*), the input resistance of the amplifier is increased. When the feedback network is connected in parallel with the *output* load (*Figure 2.16*) the effect is to reduce output

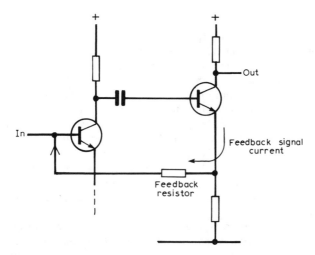

Figure 2.15. *A feedback circuit in which the feedback signal is in shunt with the input signal. At the output, the feedback resistor is connected in series with the output load*

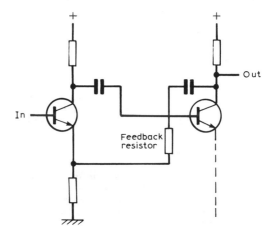

Figure 2.16. A feedback circuit in which the feedback signal is in series with the input signal through the emitter junction of the first transistor. The feedback resistor is also connected in parallel with the output load

resistance, and when the feedback is connected in series (*Figure 2.15*), the effect is to increase output resistance. The effects on output resistance are generally small compared to the effects on input resistance.

Heatsinks

A transistor passing a steady (or average) current *I* and with a steady (or average) voltage *V* between collector and emitter dissipates a power of *VI* watts. This electrical power is converted into heat at the collector-base junction, and unless this heat can be removed the temperature of the junction will rise until the junction fails irreversibly. Heat is removed in two stages, by conduction to the case or other metal work of the transistor, and into heatsinks if fitted, then by convection into the air. The temperature of the junction will stabilise when the rate of removing heat, measured in watts, is exactly equal to the electrical power dissipation; this may, however, happen only when the junction temperature is too high for continuous operation. The power dissipation of a transistor is limited therefore mainly by the rate at which heat can be removed.

For practical purposes, the resistance to heat transfer is measured by the quantity called thermal resistance, θ, whose units are $^{\circ}$C/W. The same measuring units are used for convection from heatsinks as for conduction through the transistor, so that all the figures of thermal resistance from the collector-base junction to the air can be added together as for resistor values in series. The temperature difference between the junction and the air around the heatsink is then found by multiplying the total thermal resistance by the number of watts dissipated

$$T° = \theta \times W$$

This latter figure is a temperature *difference,* so that to find the actual junction temperature, the temperature of the air around the heatsink (the ambient temperature) must be added to this figure. An ambient figure of between 30°C (for domestic equipment) and 70°C (for industrial equipment) may be taken. If this procedure ends with a calculated junction temperature higher than the manufacturer's rated values (120°C to 200°C for silicon transistors), then the dissipated power must be reduced, a larger heatsink used, or a water cooled heatsink used. Large power transistors are designed so that the transfer of heat from junction to case is efficient, with a low value of thermal resistance, and the largest thermal resistance in the 'circuit' is that of the heatsink-to-air. Small transistors generally have much higher thermal resistance values, so that the heatsinking is not so effective.

To ensure low thermal resistance, the collector of medium or high-power transistors is connected directly to the case. To prevent unintended short circuits, the heatsink may have to be insulated from other metal work, or the transistor insulated from the heatsink using mica washers. Such washers used with silicone heatsink grease can have thermal resistance values of less than 1°C/W and are available from transistor manufacturers or components specialists. The use of mica washers makes it possible to use a metal chassis as heatsink or to mount several transistors on the same heatsink.

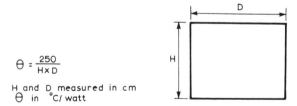

$$\theta = \frac{250}{H \times D}$$

H and D measured in cm
θ in °C/ watt

Figure 2.17. An approximate guide to the thermal resistance of a metal fin

The calculation of thermal resistance for heatsinks is not particularly simple, but *Figure 2.17* shows an approximate formula. Measurement of thermal resistance can be carried out by bolting a 25 W wirewound resistor of the metal-cased type to the heatsink. A 2.2 Ω value is suitable and will dissipate 4 W at 3 V and 16.4 W at 6 V. The temperature of the heatsink is measured when conditions have stabilised (no variation of temperature in one minute), and the electrical power figure is divided by the difference between heatsink temperature and ambient (air) temperature. This method is not precise, but gives values which are suitable for practical work.

Switching circuits

A linear amplifier circuit creates an 'enlarged' copy of a waveform. A pulse (logic) switching circuit charges rapidly from one value of voltage or current to another for a small change of voltage or current at the input. The output waveform need not be similar in shape to the input waveform, but the changes of voltage or current at the output should taken place with only a small time delay (a microsecond or less) after the changes at the input.

The bipolar (or junction) transistor has a good switching action because of its large G_m figure. A useful rule-of-thumb is that the collector current of a transistor will be increased tenfold by an extra 60 mV at the base, provided that the transistor is conducting before the extra base voltage is added, and is not saturated by the extra current. Current switches can thus be easily achieved, and a stage of current amplification can be added if larger current swings or smaller voltage swings are needed (*Figure 2.18*). A voltage switching stage must

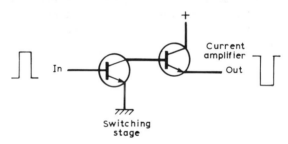

Figure 2.18. Adding a current-amplifying stage to a simple switching transistor

use some form of load to convert the current changes at the collector into voltage changes. If this load is a resistor, the switch-on of the transistor may be faster than the switch-off. Stray capacitances between the collector and earth are discharged rapidly by the current through the collector at switch-on, but must be charged through the load resistor when the transistor is switched off (*Figure 2.19*). If the rise time of the wave does not need to be short, this can be overcome by using a comparatively low value resistor (1 kΩ or less). An alternative method is shown in *Figure 2.20* using series connected transistors switching in either direction.

For fast switching, the stored charge of transistors can cause problems. During the time when the transistor is conducting, the emitter is injecting charges into the base region. These charges cannot disappear instantly when the base bias is reversed, so that the transistor will conduct momentarily in the reverse direction. As a result, the

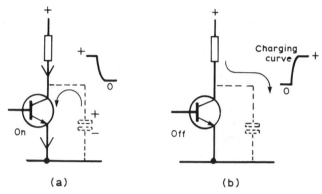

(a)   (b)

Figure 2.19. Charging and discharging stray capacitances. When the transistor conducts (a) the stray capacitance is rapidly discharged, and the voltage drop at the collector is sharp. When the transistor cuts off (b), the stray capacitance is recharged through the load resistor, causing a slower voltage rise

circuit of *Figure 2.20* can suffer from excessive dissipation at high switching speeds, since for short intervals, both transistors will be conducting. Manufacturers of switching transistors at one time quoted figures of stored charge, Q in units of picocoulombs (pC) or nanocoulombs (nC), but nowadays, generally quote the more useful turn-on

Figure 2.20. Using a two-transistor output circuit so that the switching is equally rapid in both directions. This type of output stage is used in TTL digital i.c.s

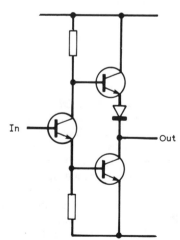

and turn-off times, in nanoseconds (ns) under specified conditions. If only stored charge figures are given, an approximate value for turn-off time can be obtained from the equation

$$t = \frac{Q}{I}$$

where t is turn-off time in nanoseconds (10^{-9}s), Q is stored charge in pC and I is the current in mA which is to be switched off. Transistor switch-off times are improved by reverse-biasing the base, but some care has to be taken not to exceed the reverse voltage limits, since the base-emitter junction will break down at moderate voltages.

A considerable improvement in switch-off times is also obtained if the transistor is not allowed to saturate during its switch-on period; this has to be done by clamping the base voltage and is not easy because of the considerable variation of switch-on voltage between one transistor and another. The fastest switching times are achieved by 'current switching' circuits, in which the transistor is never saturated nor cut-off.

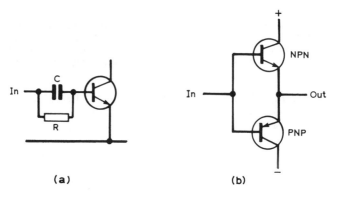

Figure 2.21. Two common switching circuit tricks. (a) use of a base-compensation capacitor, (b) using a complementary output circuit with no load resistor

Some circuits commonly used for switching circuits are shown in *Figure 2.21*, where *(a)* shows the use of a time constant *RC* in series with the base of the transistor; *C* should be adjusted for the best shape of leading and trailing edges. *Figure 2.21b* shows the familiar double-emitter-follower circuit which uses transistors both to charge and to discharge stray capacitance.

Gating of analogue signals is an action similar to that of pulse (logic) switching, but the switch may be a series component rather than a shunt component, with the added reduction that it should not distort the analogue signal while in the ON state. Diodes, bipolar transistors, and FETs can all be used in such gating circuits. *Figure 2.22* shows a

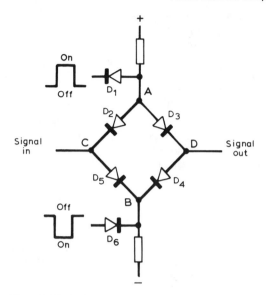

Figure 2.22. A diode-bridge gate circuit. In the off state, diodes D_1 and D_6 conduct, so that no current flows through D_2, D_3, D_4, D_5. When the gate is switched on by symmetrical pulses, D_1 and D_6 shut off, allowing the other diodes to conduct, so that the input signal can reach the output. Using symmetrical switching signals ensures that very little of the switching waveform appears in the output signal

diode bridge gate. When current flows through the diodes, assuming that there is a large resistance between point A and earth and between point B and earth, then there is a low resistance path C — D for signals in either direction. If the diodes are well matched, the d.c. level at D should be identical to that at C. Such a voltage difference is called an offset and is undesirable. When current ceases to flow, the diodes

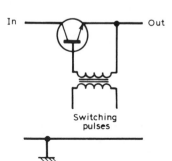

Figure 2.23. A gate circuit using a single transistor

become non-conducting, the gate closes. Offset voltages may be around 10 mV (comparatively high), but very high operating speeds are possible when switching diodes are used.

Bipolar transistors can be used as gates, with the offset voltage between collector and emitter being lowest (less than 2 nV) for a saturated transistor when the normal collector and emitter terminals are *reversed*. As a series switch, however, the transistor suffers from the requirement of a switching pulse applied between base and emitter, so that a transformer must be used to apply the switching voltage if a single transistor is to be used (*Figure 2.23*). A very common gating circuit is the long-tailed pair of *Figure 2.24*, which is, however, useful

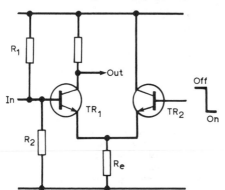

Figure 2.24. The long-tailed pair gate. When Tr_2 is switched off Tr_1 is normally biased by R_1, R_2, and acts as an inverting amplifier. When Tr_2 is switched on, with its base voltage several volts higher than the normal bias voltage of Tr_1, then Tr_1 is biased off

only when the offset voltages (v_{be}) and the voltage change caused by switching are unimportant.

FETs may be used as shunt or series switches (*Figure 2.25*). The shunt switch is considerably easier to drive because the source terminal is earthed. Offset voltages of less than 10 μV are obtainable, and practically no drive current to the gate is needed. The disadvantage of the FET is that its resistance when switched ON is much higher (up to 1 kΩ) than that of a bipolar transistor.

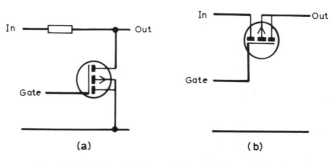

Figure 2.25. Using FETs as switches: (a) shunt, (b) series

Other switching devices

Unijunctions have two base contacts and an emitter contact, forming a device with a single junction which does not conduct until the voltage between the emitter and base contact 1 (*Figure 2.26*) reaches a specified level. At this level, the whole device becomes conductive. The unijunction

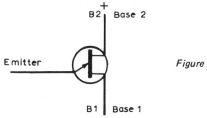

Figure 2.26. Unijunction symbol

is used to generate short pulses, using circuits such as that of *Figure 2.27*. The frequencies of operation of this circuit are not noticeably affected by supply voltage changes, since the unijunction fires (becomes conductive) at a definite fraction of the supply voltage. The intrinsic stand-off ratio, n, is defined as

$$\frac{\text{firing voltage } (e - b_1)}{\text{supply voltage } (b_2 - b_1)}$$

and has values ranging typically from 0.5 to 0.86. Pulse repetition rates up to 1 MHz are obtainable.

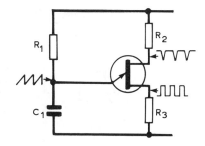

Figure 2.27. A unijunction oscillator. R_2, R_3 are about 100 ohms each, and the frequency of oscillation is determined by the time constant RC

Programmable unijunction transistors (PUTs) have three terminals, one of which is used to set the value of intrinsic stand-off ratio, n, by connection to a potential divider (*Figure 2.28*). Firing will occur at the programmed voltage; the frequency range is generally up to 10 kHz.

Thyristors are controlled silicon diodes which are switched into conduction by a brief pulse or a steady voltage at the gate terminal. Voltage of 0.8 to 1.5 V and currents of a few μA up to 30 mA are needed at the gate, according to the current rating of the thyristor.

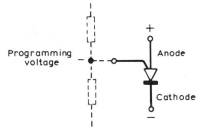

Figure 2.28. A programmable unijunction transistor (PUT). The firing voltage between anode and cathode is selected by the voltage applied to the third electrode

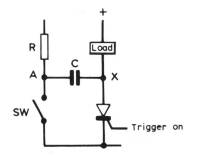

Figure 2.29. A capacitor turn-off circuit for a thyristor. When the switch is momentarily closed, the sudden voltage drop at A will cause an equal drop at X, turning off the thyristor until it is triggered again

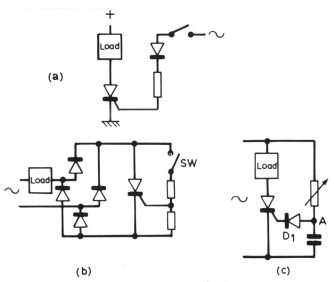

Figure 2.30. A.C. thyristor circuits. (a) Basic half-wave a.c. relay circuit. (b) A full-wave relay circuit. (c) Basic phase-control circuit. In the half-cycle during which the thyristor can conduct, the gate is activated only when the voltage at A has risen enough to cause the trigger diode D_1 to conduct. The time in the cycle at which conduction starts is controlled by the setting of the variable resistor

The thyristor ceases to conduct only when the voltage between anode and cathode falls to a low value (about 0.2 V) or when the current between anode and cathode becomes very low (less than 1 mA). D.C. switching circuits need some form of capacitor discharge circuit (*Figure 2.29*) to switch off the load. A.C. switching circuits, using a.c. or full wave rectified waveforms, are switched off by the waveform itself on each cycle. A few typical a.c. thyristor circuits are shown in *Figure 2.30*. Note that the gate signal may have to be applied through a pulse transformer, particularly when the thyristor switches mains currents, to avoid connecting the firing circuits to the gate.

Triacs are two-way thyristors whose terminals are labelled MT1, MT2 and gate. For reliable firing, the pulse at the gate should be of the same polarity as MT2 (some circuits are shown in *Figure 2.31*).

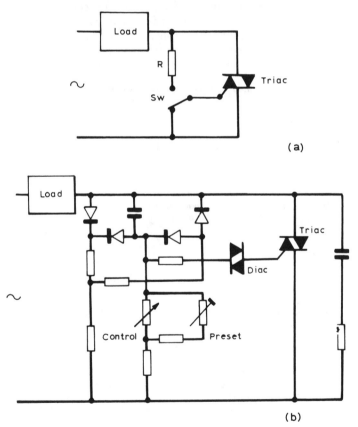

Figure 2.31. Triac circuits. (a) Basic full-wave relay circuit. (b) Power regulator circuit, using a diac trigger diode, and radio interference suppression circuit across the triac

Firing pulses for thyristors and triacs can be obtained from uni-junctions or from other types of trigger device such as diacs, silicon bidirectional switches, four-layer diodes or silicon unidirectional switches. The diac, or bidirectional trigger diode, is non-conductive in either direction until its breakdown voltage is exceeded, after which the device conducts readily until the voltage across its terminals (either polarity) is low. Firing voltages of 20 to 36 V are typical, and the 'breakback' voltage (at which the device ceases to conduct) is typically 6 V. Brief peak currents of 2 A are possible. The silicon bidirectional switch also uses a gate electrode, but operates with one polarity only. Four-layer diodes have lower firing and breakback voltages than diodes, but essentially similar characteristics.

The silicon controlled switch (SCS) is a useful device with four electrodes which can be used, according to connections, either as a programmable unijunction or as a low-power thyristor. The connections are referred to as anode, cathode, gate-anode and gate-cathode. If the gate-cathode is used together with the anode and cathode, thyristor operation (at low currents) is obtained; if the gate-anode is used, the

Table 2.1 PRO-ELECTRON CODING

The first letter indicates the semiconductor material used:

A	Germanium
B	Silicon
C	Gallium arsenide and similar compounds
D	Indium antimonide and similar compounds
R	Cadmium sulphide and similar compounds

The second letter indicates the application of the device:

A	Detector diode, high speed diode, mixer diode
B	Variable capacitance (varicap) diode
C	A.F. (not power) transistor
D	A.F. power transistor
E	Tunnel diode
F	R.F. (not power) transistor
G	Miscellaneous
L	R.F. power transistor
N	Photocoupler
P	Radiation detector (photodiode, phototransistor, etc.)
Q	Radiation generator (LED etc.)
R	Control and switching device (such as thyristor)
S	Switching transistor, low power
T	Control and switching device (such as a triac)
U	Switching transistor, high power
X	Multiplier diode (varactor or step diode)
Y	Rectifier, booster or efficiency diode
Z	Voltage reference (zener), regulator or transient suppressor diode.

The remainder of the code is a serial number. For consumer applications, such as radio, TV, hi-fi, this has three figures. For industrial and telecommunications use, a letter W, X, Y or Z and two figures are used.

device behaves as a PUT. The unused electrode is generally left open circuit.

Table 2.1 shows the European Pro-Electron coding used for semi-conductors. The US (JEDEC) 1N and 2N numbers are registration numbers only (this also applies to the Japanese 2SA, 2SB etc. system), and the function of a semiconductor cannot be guessed from the number.

Chapter 3

Discrete Component Circuits

This Chapter illustrates a selection of well established circuits and data, and comments are reduced to a minimum so as to include the greatest possible number of useful circuits. The common-emitter and a few other amplifier circuits have already been dealt with in Chapter 2.

Where several different types of circuits are shown (as for oscillators) practical considerations may dictate the choice of design. For example,

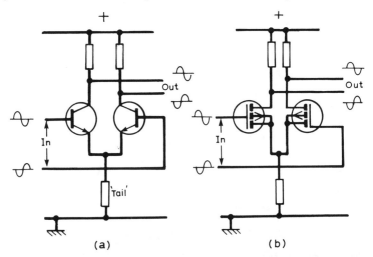

Figure 3.1. The long-tailed pair circuit using (a) bipolar transistors, (b) p-channel MOSFETS. Balanced input signals, as shown, are amplified, but unbalanced signals (in the same phase at each input) are attenuated

a Hartley oscillator uses a tapped coil, but the arrangements for frequency variation may be more convenient than those for a Colpitts oscillator, which uses a capacitive tapping. Some crystal oscillator circuits are not always self-starting, particularly with 'difficult' crystals. For this reason, as many variations on basic circuits have been shown as is feasible within the space.

The long-tailed pair, shown both in bipolar and in FET form in *Figure 3.1* is the most versatile of all transistor circuits, which is why it is so extensively used for linear i.c.s. A common-mode signal is a signal applied in the same phase to both bases or gates. Any amplification of such a common mode signal can only be caused by lack of balance between the transistors, so that this value of gain is low. The difference signal is amplified with a considerably greater gain. The long-tailed pair is most effective when used as a balanced amplifier, with balanced input and output, but single-ended inputs or outputs can be provided, as shown in *Figure 3.2a* and *b*. The overall voltage gain of a long-tailed pair circuit is about half the gain that would be obtained from one of the transistors, using the same load and bias conditions.

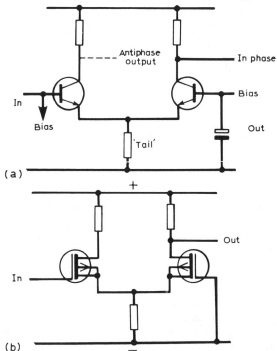

Figure 3.2. Single-ended inputs and outputs on a long-tailed pair circuit. The second input is earthed to signals — no bias arrangements have been shown. (a) Bipolar transistors, (b) p-channel MOSFETs

Figure 3.3 ilustrates a magnetic pickup preamplifier circuit. The problem here is to apply the frequency correction (equalisation) needed for disc replay. Discs are recorded to the RIAA standard (see BS 1928: 1965), in which bass frequencies are attenuated (to prevent excessive cutter movement) and treble frequencies are boosted (to have an amplitude well above that of surface noise). This deliberate distortion must be corrected at playback by *CR* networks; to achieve this correctly, three time constants, 75 μs, 318 μs and 3180 μs, are used. The 75 μs time constant is of a treble cut filter, taking effect (that is, with its turnover point) at 2.12 kHz; the 318 μs kHz is a bass boost starting at 500 Hz, and the 3180 μs time constant is a final stage of bass cut at 50 Hz and under.

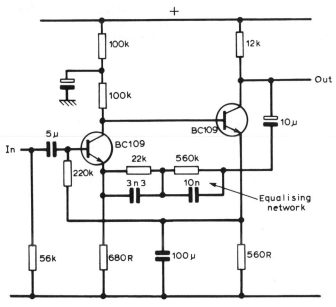

Figure 3.3. An input stage for magnetic pickup cartridges. This stage combines high gain with equalisation, and a fairly high voltage supply (40 V or more) is needed to avoid overloading on transients

The output from most magnetic cartridges, other than the moving-coil type, is around 5 mV at the standard conditions of 5 cm/s stylus velocity and 1 kHz signal. This corresponds to a signal amplitude of close to the minimum most amplifiers need to give full output for 2.5 mV input at full volume setting. The preamplifier should present an input resistance of around 50 kΩ and should be capable of accepting considerable overloads at the input, 50 mV or more, without noticeable distortion. The trend at the time of writing is away from the feedback-loop type of equalisation, and towards an equalising network of purely passive type applied after a 'flat' preamplifier stage — this is held to be advantageous because of the effects of transients on feedback amplifiers, particularly when momentarily overloaded.

Figure 3.4 shows some tape/cassette input circuits. Once again, equalisation is needed, but the time constants are not quite so universally agreed; they appear to change almost annually as new tape materials and new types of tape head construction appear. In addition to these 'standard' corrections, individual tape decks may need further corrections, a multiplex filter may be included to remove f.m. stereo subcarrier

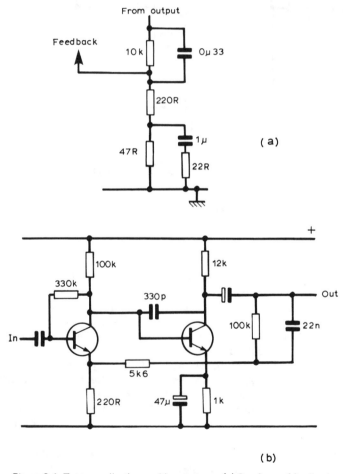

Figure 3.4. Tape equalisation and input stages. (a) One form of feedback network which can be used for reel-to-reel tape equalisation. (b) A cassette recorder input stage, using rather different time constants in the equalisation networks

signals, and noise-cancelling circuits, such as the Dolby circuits, may be
used. At the last count, the equalisation frequencies being used on
replay were 3180 μs for all tapes and either 70 μs or 120 μs for
chrome and ferric tapes respectively; ferrichrome and pure iron particle
tapes are replayed at 70 μs. The equalisation needed for recording
amplifiers is too specialised to include here partly because recording
equalisation time constants depend much more on individual needs.

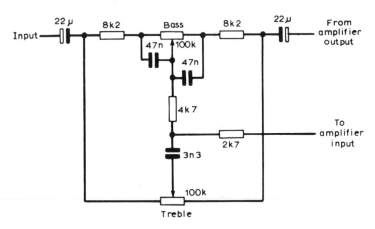

*Figute 3.5. A Baxandall type of tone control circuit. This circuit is normally
located between two voltage amplifier transistors which provide the
necessary gain*

Figure 3.5 shows a version of the Baxandall tone control circuit,
which is virtually the standard method of tone control used nowadays.
There is very little interaction between the treble and the bass controls,
low distortion, and a good range of control; about 20 dB of boost or
cut. The Baxandall circuit is usually located between bipolar transistors,
but several designs claim significant improvements by using a FET at
the output of the control stage.

Figure 3.6 deals with active filters. These designs use only resistors
and capacitors, together with semiconductors, and are considerably
simpler to design than *LC* filters. Low-pass, high-pass, bandpass and
notch filters are illustrated in *Figure 3.6*. The filters generally have a
slope of 12 dB per octave (meaning that the response changes by 12 dB
for each doubling or halving of frequency).

Figure 3.7 shows a typical input preamplifier stage for a moving-coil
microphone. The particular features here are low noise operation,
matching a fairly low input resistance, high gain, and hum rejection.
The low output and low resistance of the moving-coil microphone
requires the use of a microphone transformers. If a balanced layout is
possible, hum pickup can be greatly reduced.

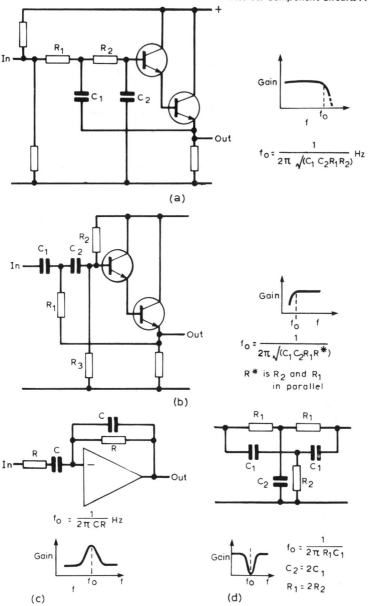

Figure 3.6. Active filter circuits. (a) Low pass, (b) High pass. These two types are Sallen & Key filters, named after the inventors. Circuit (c) is a bandpass filter using the Wien bridge network and an inverting amplifier. The gain of the amplifier determines the effective Q of the circuit, so that higher gain causes narrower bandwidth. Circuit (d) is a passive twin-T notch filter. The selectivity can be improved by using positive feedback

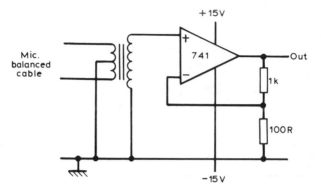

Figure 3.7. A moving-coil microphone input circuit. A suitable
input transformer is essential if high-quality results are expected.
The transformer should preferably be supplied by the makers of
the microphone

The next three sets of circuits deal with audio output stages. Class A
stages are those in which the transistor(s) are always biased on and
never saturated (bottomed). A Class A stage may use a single transistor
(a single-ended stage) or two transistors which share the current in some
way (a push-pull stage), but the efficiency is low. % Efficiency is
defined as

$$\frac{\text{power dissipated in the load} \times 100}{\text{total power dissipated in the output stage}}$$

and is always less than 50% for Class A operation.

A Class A stage should pass the same current when no signal is applied
as when maximum signal is applied. Because of this, the dissipation is
large, so that large-area heatsinks are needed for the output transistors.

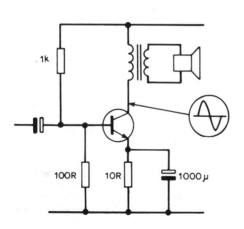

Figure 3.8. A Class A
single-ended output stage.
Good heatsinking is es-
sential

Class B audio operation uses two (or more) transistors biased so that one conducts on one half of the waveform and the other on the remaining half. Some bias must be applied to avoid 'crossover distortion' due to the range of base-emitter voltage for which neither transistor would conduct in the absence of bias. Class B audio stages can have efficiency figures as high as 75%, though at the expense of rather higher distortion than a Class A stage using the same layout. The higher efficiency enables greater output power to be obtained with smaller heatsinks, and the use of negative feedback can, with careful design, reduce distortion to negligible levels.

Figure 3.8 shows a Class A single-ended power output stage, suitable for general-purpose use such as car radio operation. *Figure 3.9* shows the totem-pole or single-ended push-pull circuit, which can be used for either Class A or Class B operation according to the bias level. This version uses complementary symmetry — the output transistors are

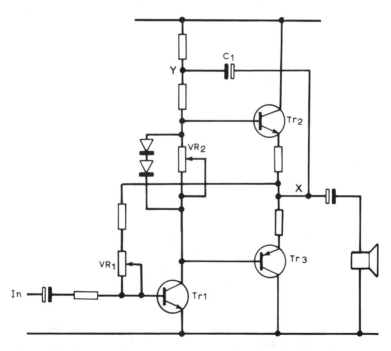

Figure 3.9. A single-ended push-pull (totem-pole) Class B output stage, using complementary power transistors. VR$_1$ sets the voltage at point X to half of the supply voltage, VR$_2$ sets the quiescent (no signal) current through the output transistors. C$_1$ is a 'bootstrap' capacitor which feeds back in-phase signals to point Y, increasing input impedance. Oscillation is avoided because the gain of Tr$_2$ is less than 1

pnp and npn types. When complementary output transistors cannot be obtained, a pseudo-complementary circuit, such as that of *Figure 3.10*, can be used, though this is not truly symmetrical.

Figures 3.11, 3.12 and *3.13* illustrate some of the circuits used for wideband voltage amplification. *Figure 3.11* deals with methods of frequency compensation using inductors or capacitors to compensate for the shunting effect of stray capacitances. *Figure 3.12* shows a circuit which uses feedback to reduce the gain and so extend the flat portion

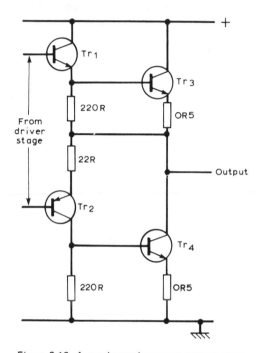

Figure 3.10. A quasi-complementary output stage. The low power complementary transistors Tr_1, Tr_2 drive the high-power output pair, Tr_3, TR_4, which are not complementary types. The circuit is not symmetrical, and can cause considerable distortion when overdriven

of the frequency range — this is a useful basic circuit for video frequencies. *Figure 3.13* shows a cascode amplifier, a type of construction which was widely used in the days of valves, but which has been strangely neglected in transistor circuits. The advantage of the cascode is stability, because there is practically no feedback from output to input and high gain over a large bandwidth. FET cascodes and combinations of FET and bipolar transistors can also be used.

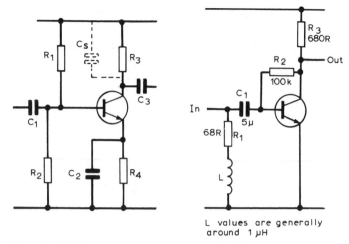

Figure 3.11. Frequency compensation for wideband amplifiers. (a) Capacitive compensation. The value of C_2 is chosen so that R_4 is progressively decoupled at high frequencies. As a rough guide, $C_s R_3$ should equal $C_2 R_4$. (b) Inductive shunt compensation. The value of L is chosen so as to resonate with the input capacitance of the transistor at a frequency above that of the uncompensated 3 dB point. These compensation methods are useful, but cannot compensate for low gain caused by an unsuitable transistor type. Transistors capable of amplification at high frequencies must be used in these circuits

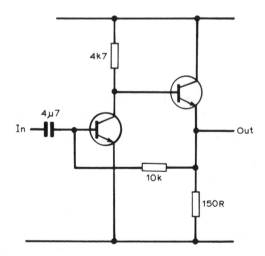

Figure 3.12. A feedback pair circuit which is capable of wideband amplification when suitable transistors are used. Bandwidths of 5 MHz or more are obtainable

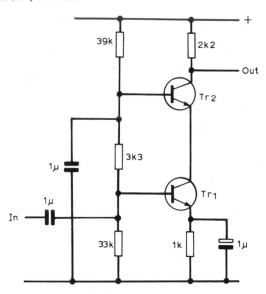

Figure 3.13. The cascode circuit, using, in this example, bipolar transistors. This is a exceptionally stable circuit, because of the isolation between input and output, and has high gain and a large bandwidth

The circuits of *Figures 3.14* to *3.16* are of sine wave oscillators which operate at radio frequencies. The Hartley type of oscillator (*Figure 3.14*) uses a tapped coil; the Colpitts type (*Figure 3.15*) uses a capacitor tap. Though these are not the only r.f. oscillator circuits, they are the circuits most commonly used for variable frequency oscillators. *Figure 3.16* illustrates some crystal controlled oscillator circuits. The frequency of the output need not be the fundamental crystal frequency, since most crystals will oscillate at higher harmonics (overtones) and harmonics can be selected at the output. Frequency multiplier stages (see *Figure 3.26*) can then be used to obtain still higher frequencies.

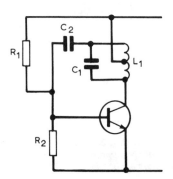

Figure 3.14. The Hartley oscillator. The resonant circuit is L_1C_1, and the value of C_2 should be chosen so that the amount of positive feedback is not excessive, since this causes a distorted waveform. R_1 should be chosen so that the transistor is just drawing current when C_1 is short-circuited

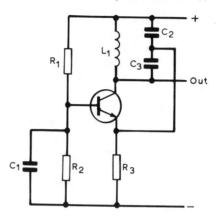

Figure 3.15. The Colpitts oscillator. The tapping is provided by C_2, C_3, and it is the series combination of these two capacitors which tunes L_1. As a rough guide, the value of C_3 should be about ten times the value of C_2 to avoid over-driving. R_1 is chosen to give about 1 mA of collector current. C_1 must not be omitted, because oscillation is impossible (because of negative feedback to the base) unless the base is decoupled

For low frequencies, oscillators such as the Wien bridge (*Figure 3.17*) or twin-T types are extensively used. Usable frequency ranges are from 1 Hz, or lower, to around 1 MHz.

Untuned or aperiodic oscillators are important as generators of square and pulse waveforms. *Figure 3.18* shows the familiar multivibrator (astable) together with modifications which improve the shape of the waveform. The less familiar serial multivibrator is shown in *Figure 3.19;* this circuit is a useful source of narrow pulses. When a pulse of a determined, or variable, width is required from any input (trigger)

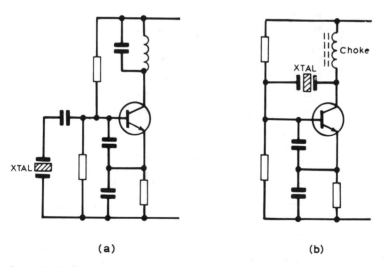

(a)   (b)

Figure 3.16. Two versions of a Colpitts crystal oscillator, showing alternative positions for the cystal

pulse, a monostable circuit must be used. *Figure 3.20* shows a monostable circuit, with a block diagram to illustrate how a combination
of astable and monostable can form a useful pulse generator.

Figure 3.21 shows the basic bistable circuit, now rather a rarity in
the discrete form thanks to the low price of i.c. versions. The Schmitt
trigger is illustrated in *Figure 3.22*; its utility is as a comparator and

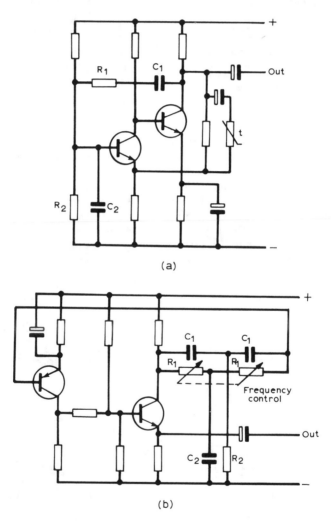

Figure 3.17. Low-frequency oscillators. (a) Wien bridge, (b) twin-T.
The Wien bridge circuit uses a thermistor to keep the amplitude of
the output signal constant. R_1, R_2 may be a ganged variable if a
variable frequency output is needed

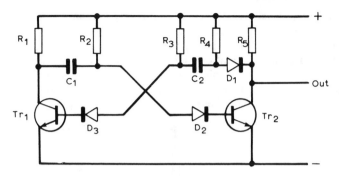

$$T_1 = 0.7\, C_1\, R_2$$
$$T_2 = 0.7\, C_2\, R_3$$
$$f = \frac{1}{T_1 + T_2}$$

Figure 3.18. The astable multivibrator. The frequency of operation is given by the formula shown. The diodes D_2, D_3 prevent breakdown of the base-emitter junctions of the transistors when the transistors are turned off, and D_1 isolates the collector of Tr_2 from C_2 when Tr_2 switches off. In this way, a fast-rising waveform can be obtained

trigger stage which gives a sharply changing output from a slowly changing input. The hysteresis (voltage difference between the switching points) is a particularly valuable feature of this circuit. A circuit with hysteresis will switch positively in each direction with no tendency to 'flutter' or oscillation, so that Schmitt trigger circuits are used extensively where electronic sensors have replaced purely mechanical devices such as thermostats.

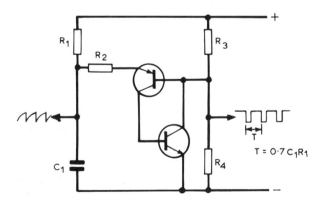

Figure 3.19. The serial astable. Only one time constant is needed, and the outputs are a sawtooth and a pulse as shown. Usually $R_3 = R_4$; values of around 10 kΩ are usual

Radio-frequency circuits are represented here by only a few general examples, because the circuits and design methods that have to be used are fairly specialised, particularly for transmission, and the reader who wishes more information on purely r.f. circuits is referred to the excellent amateur radio publications. *Figure 3.23* shows an a.m./f.m. i.f. amplifier for 470 kHz and 10.7 MHz such as would be used in a.m./ f.m. receivers. This design uses a common emitter amplifier, since the

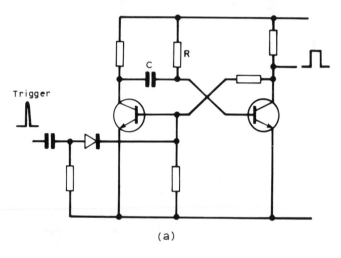

(a)

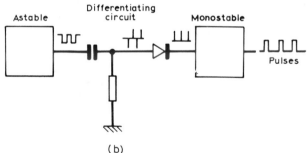

(b)

Figure 3.20. The monostable (a). The pulse width of the output pulse is determined by the time constant CR. The block diagram (b) shows how an astable (to determine frequency) and a monostable (to determine pulse width) can be combined to form a precise pulse generator

operating frequency is well below the turnover frequency, f_r (at which gain is unity) for the transistor. For v.h.f. use, common base stages (*Figure 3.24*) and dual-gate FETs (*Figure 3.25*) are extensively used. *Figure 3.26* shows frequency multiplier and intermediate stages for transmitters, and *Figure 3.27* a selection of low-pair output (power

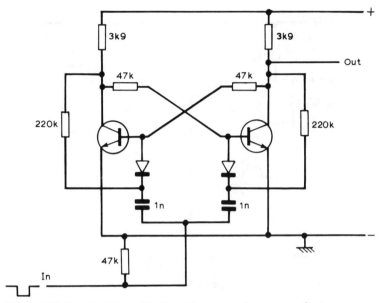

Figure 3.21. The bistable, or flip-flop. The output changes state (high to low or low to high) at each complete input pulse

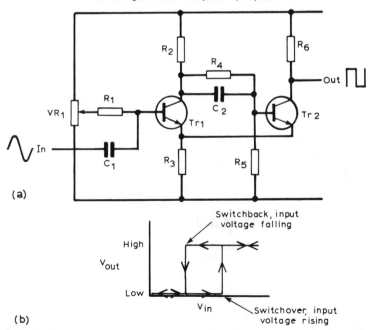

Figure 3.22. The Schmitt trigger circuit. (a) This is a switching circuit whose output is always flat-topped and steep-sided whatever the input waveform. The characteristic (b) shows hysteresis — a difference between the switching voltages depending on the direction of change of the input voltage

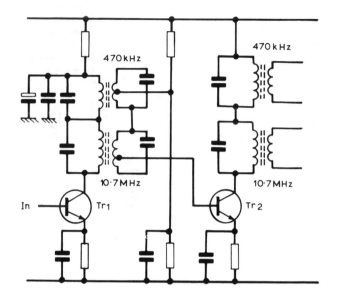

Figure 3.23. An i.f. amplifier typical of a.m./f.m. receivers. The frequency difference between the i.f.s is so great that no special filtering is needed, but the 10.7 MHz transformers must be located close to their transistors

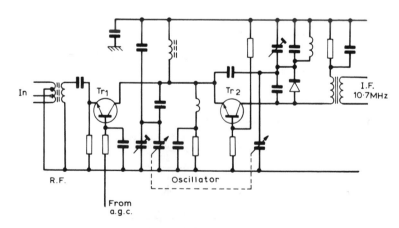

Figure 3.24. A typical f.m. 'front-end', using a common-base r.f. amplifier and a common-base oscillator. All transistors will operate at higher frequencies in the common-base connection than in the common-emitter connection

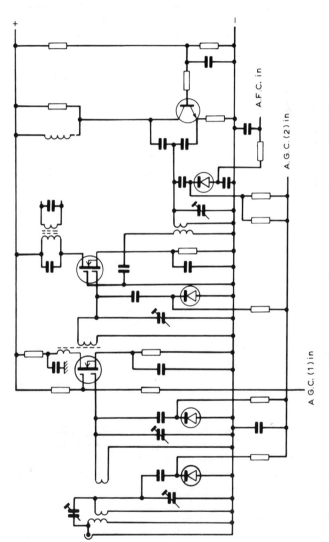

Figure 3.25. An f.m. front end of a high-quality tuner (Armstrong), using dual-gate MOSFETs. Using both gates enables the r.f. a.g.c. to be completely isolated from the signal input, and the mixer signal input to be isolated from the oscillator input

A.F.C. in

A.G.C.(2) in

A.G.C.(1) in

amplifier or p.a.) stages. Transmitters which use variable frequency oscillators (v.f.o.) will require broadband output stages as distinct from sharply tuned stages, and this precludes the use of Class C amplifiers (in which the transistor conducts only on signal peaks). Without a sharply tuned, high Q, load, Class C operation introduces too much distortion (causing unwanted harmonics) and so Class B is preferable.

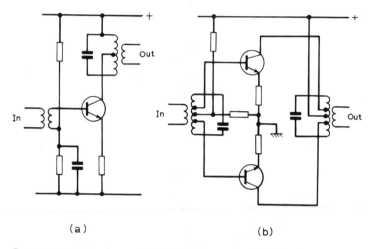

(a)   (b)

Figure 3.26. Frequency multipliers for transmitters. (a) A single transistor multiplier for even or odd multiples, (b) a push-pull multiplier for odd multiples. In each type of circuit, the output is tuned to a frequency which is a multiple of the input frequency. Other techniques not shown here include push-push, in which two transistors have antiphase signals at their collectors, but both feed the same output at the collectors. This circuit is used for even multiples. Varactor diodes are also extensively used as multipliers at low power levels

To carry information by radio or by digital signals requires some form of modulation and demodulation. For radio use amplitude modulation and frequency modulation are the most common techniques. Straightforward amplitude modulation produces two sidebands, with only one third of the total power in the sidebands, so that double sideband a.m. is used virtually only for medium and long wave broadcasting. Short wave communications use various forms of single sideband or suppressed carrier a.m. systems; v.h.f. radio broadcasting uses wideband f.m. and other v.h.f. communications use narrow-band f.m.

Figure 3.28 shows two simple modulator circuits, excluding specialised types. Carrier suppression can be achieved by balanced modulators in which the bridge circuit enables the carrier frequency to be balanced out while leaving sideband frequencies unaffected.

Sideband removal can be achieved by crystal filters, a fairly straight-forward technique which is applicable only if the transmitting frequency is fixed, or by a phase-shift modulator which makes use of the phase shift which occurs during modulation. Frequency modulation, unlike amplitude modulation, is carried out on the oscillator itself, so requiring

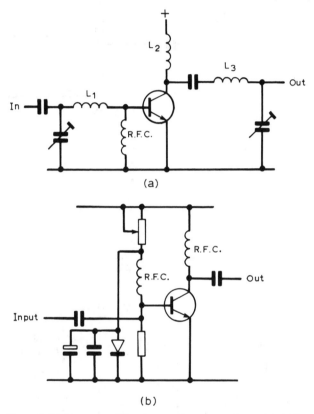

Figure 3.27. Power amplifiers for transistor transmitters. (a) A Class C single transistor p.a. stage, (b) a Class B design, necessary for single-sideband transmitters. Tuning inductors have been omitted for clarity. In both circuits some decoupling capacitors have not been shown — complete decoupling is essential. At the higher frequencies, circuit layout is critical, and the circuit diagram becomes less important than the physical layout

reasonably linear operation of the stages following the oscillator. *Figure 3.29* illustrates some types of discrete component demodulators.

Pulse modulation systems are used extensively in applications ranging from data processing to radar. Pulse amplitude modulation and frequency modulation is essentially similar in nature to a.m. and f.m.

of sine waves, and will not be considered here. Forms of modulation peculiar to pulse operation are pulse width modulation (p.w.m.), pulse position modulation (p.p.m.) and pulse-code modulation (p.c.m.). A technique which is not a pulse modulation system but which is

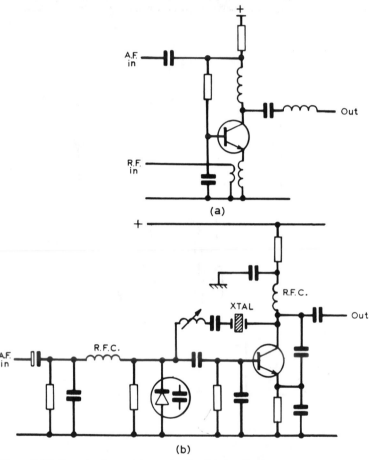

Figure 3.28. Two simple modulator circuits. (a) A collector modulated stage for an a.m. transmitter, (b) a varactor diode f.m. modulator

extensively used for coding slow pulse information is frequency shift keying (f.s.k.) in which the high (logic 1) and low (logic 0) voltages of a pulse are represented by different audio frequencies.

Figure 3.30 is concerned with optical circuits, including LED devices and light detectors.

The circuits of *Figure 3.31* deal with power supply units. *Figure 3.31* shows the no-load voltage output, and the relationship between d.c. load voltage, minimum voltage, and a.c. voltage at the transformer. Only capacitive input circuits have been shown, since choke-input filters are by now rather rare. The relationship between the size of the reservoir capacitor and the peak-to-peak ripple voltage is given approximately by

$$V = \frac{I_{dc} \times t}{C}$$

with I_{dc} equal to load current (amperes), t in seconds the time between voltage peaks and C the reservoir capacitance in farads.

A more convenient set of units is I_{dc} in mA, t in ms, and C in μF, using the formula unchanged. V is then the peak-to-peak ripple voltage in volts.

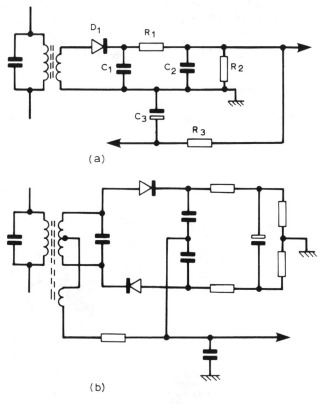

(a)

(b)

Figure 3.29. Demodulators. The a.m. demodulator (a) uses a single diode. The time constant of C_1 with $R_1 + R_2$ must be long compared with the time of a carrier wave, but short compared with the time of the highest-frequency audio wave. The f.m. demodulator (b) is a ratio detector

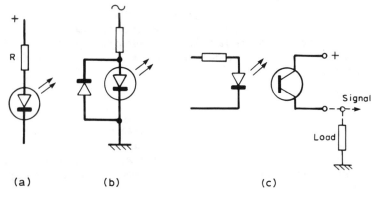

(a)   (b)   (c)

Figure 3.30. Optoelectronic circuits. (a) Driving a single LED — a current-limiting resistor must be used. (b) When an LED is operated from a.c., a diode must always be included to protect the LED from reverse voltage. (c) Optocoupler, used to couple signals at very different d.c. levels. This is useful for triac firing, or for modulating the grid of a c.r.t., since d.c. signals can be transferred, which is not possible using a transformer

All power supplies which use the simple transformer-rectifier-capacitor circuit will provide an unstabilised output, meaning that the output voltage will be affected by fluctuations in the mains voltage level and also by changes in the current drawn by the load. The internal resistance of the power supply unit causes the second effect and can be the reason for instability in amplifier circuits, or of misfiring of pulse circuits. A stabiliser circuit provides an output which, ideally, remains constant despite any reasonable fluctuation in the mains voltage and

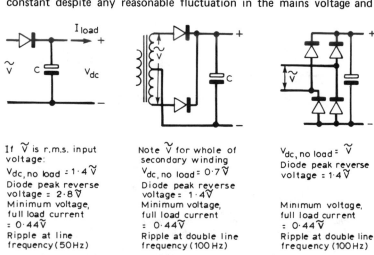

If $\tilde{V}$ is r.m.s. input voltage:
V_{dc}, no load $= 1 \cdot 4 \tilde{V}$
Diode peak reverse voltage $= 2 \cdot 8 \tilde{V}$
Minimum voltage, full load current $= 0 \cdot 44 \tilde{V}$
Ripple at line frequency (50 Hz)

Note $\tilde{V}$ for whole of secondary winding
V_{dc}, no load $= 0 \cdot 7 \tilde{V}$
Diode peak reverse voltage $= 1 \cdot 4 \tilde{V}$
Minimum voltage, full load current $= 0 \cdot 44 \tilde{V}$
Ripple at double line frequency (100 Hz)

V_{dc}, no load $= \tilde{V}$
Diode peak reverse voltage $= 1 \cdot 4 \tilde{V}$

Minimum voltage, full load current $= 0 \cdot 44 \tilde{V}$
Ripple at double line frequency (100 Hz)

Figure 3.31. Rectifier circuits in detail

has zero internal resistance so that the output voltage is unaffected by the load current.

Stabilisation is achieved by feeding into the stabiliser circuit a voltage which is higher than the planned output voltage even at the worst combination of circumstances — low mains voltage and maximum load current. The stabiliser then controls the voltage difference between input and output so that the output voltage is steady.

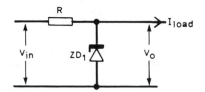

Design data: Allow 2mA minimum current through zener diode

Maximum current $= (I_{load} + 2)$mA

Diode dissipation, max. $= V_0 (I_{load} + 2)$mW

Resistor dissipation, max $=$
$(V_{in} - V_0)(I_{load} + 2)$ mW

Figure 3.32. A simple zener-diode stabiliser

Figure 3.32 shows a simple zener diode stabiliser suitable for small scale circuits taking only a few milliamps. This is a shunt stabilising circuit, so called because the stabiliser (the zener diode) is in parallel (shunt) with the load. The value of the resistor *R* is such that there will be a 'holding' current of 2 mA flowing into the zener diode even at the lowest input voltage and maximum signal current. The circuit of *Figure 3.33* (sometimes known as the 'amplified zener') is a shunt stabiliser which does not depend on dissipating power in the zener when the load current drops.

Figure 3.33. An 'amplified-zener' or shunt-regulator circuit. The transistor dissipation is greatest when the load current is least

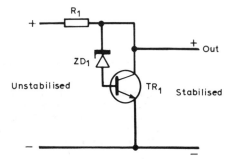

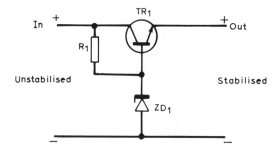

Figure 3.34. A simple series stabiliser. The transistor dissipation is greatest when the load current is maximum

Figure 3.34 shows a simple series stabiliser, using a zener diode to set the voltage at the base of an emitter follower. A more elaborate negative feedback circuit is shown in *Figure 3.35,* with provision for altering the stabilised voltage. The circuit of *Figure 3.36* also provides automatic shut-off (with the shut-off current alterable by setting the

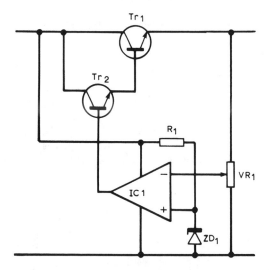

Figure 3.35. A variable-voltage series stabilised supply circuit. Tr₁ is usually a 2N3055, and Tr₂ a lower power general purpose transistor. The most suitable i.c. is the LM3900, because it continues to amplify even when the output voltage is close to either supply voltage, unlike the 741

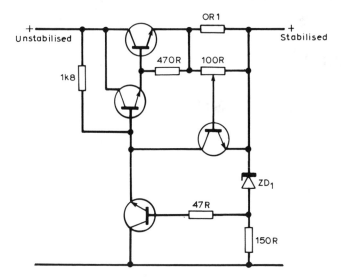

Figure. 3.36. A fixed-voltage stabiliser with over-current protection, limiting the current to an amount which is set by the potentiometer

100 Ω potentiometer) when excessive load current flows. More elaborate circuits are not considered here because of the extensive use of i.c. regulators (see Chapter 4).

Chapter 4

Linear I.C.s

Linear i.c.s are single-chip arrangements of amplifier circuits that are intended to be biased and operated in a linear way. This definition is usually extended to include i.c.s which have a comparatively slow switching action, such as the 555 timer.

The most important class of linear amplifier i.c. is the operational amplifier which features high gain, high input resistance, low output resistance and a narrow bandwidth extending to d.c. Such amplifiers are almost invariably used in negative feedback circuits, and make use of a balanced form of internal circuit (*Figure 3.1*) so that power supply hum and noise picked up by stray capacitance are both discriminated against.

The 741 is typical of operational amplifiers generally, so that the design methods, circuits and bias arrangements which are used for this i.c. can be used, with small modifications, for other types. Referring to the pinout diagram and symbol of *Figure 4.1*, the 741 uses two inputs marked + and −. These signs refer to the phase of the output signal relative to each input, so that feedback directly from the output to the + input is positive, and feedback directly from the output to the − input is negative.

The circuit arrangement of the 741 is such that, using balanced power supplies, the d.c. level at the output ought to be at zero volts when both inputs are connected to zero volts. This does not generally happen because of slight differences in internal components, so that an *input offset voltage* is needed to restore the output to zero voltage. Alternatively, the offset can be balanced out by a potentiometer connected as shown in *Figure 4.2*. Once set in this way so that the

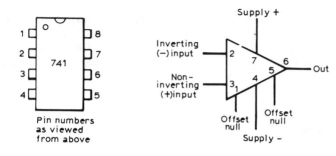

Figure 4.1. 741 operational amplifier outline, with pin numbering (a) and the connections (b). The offset-null pins are used only for d.c. amplifier applications

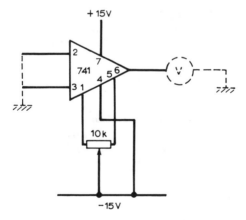

Figure 4.2. Using an offset-null control. With the inputs both earthed (balanced power supplies) and a voltmeter connected to the output (dotted lines), the 10 KΩ potentiometer is adjusted so that the output voltage is zero

output is at zero volts (with the inputs earthed), the output voltage will then slowly change (drift). The drift may be caused by temperature changes, by supply voltage changes, or simply by old age. Drift is a problem which mainly affects high-gain d.c. coupled amplifiers and long time-constant integrators; a.c. amplifier circuits, and circuits which can use d.c. feedback bias are not affected by drift.

Bias Methods

For linear amplification, both inputs must be biased to a voltage which lies approximately halfway between the supply voltages. The output voltage can then be set to the same value by:

(a)   making use of an offset-balancing potentiometer, or,

(b)   connecting the output to the (−) input through a resistor, so making use of d.c. feedback.

Method (a) is seldom used, and the use of d.c. feedback is closely tied up with the use of a.c. feedback; the two will be considered together.

The power supply may be of the balanced type, such as the ±15 V supply, or unbalanced, provided that the bias voltage of input and output is set about midway between the limits (+15 and −15, or +V and 0) of supply voltages. Bias voltages should not be set within three volts of supply voltage limits, so that when a ±15 V supply is used, the input or output voltages should not exceed +12 V or −12 V. This limitation applies to bias (steady) voltage or to instantaneous voltages. If a single-ended 24 V power supply is used, the input and output voltages should not fall below 3 V nor rise above 21 V. Beyond these limits, the amplifying action may suddenly collapse because there is not sufficient bias internally.

Basic circuits

Figure 4.3 shows the circuits for an inverting amplifier, using either balanced or unbalanced power supplies. The d.c. bias conditions are set by connecting the (+) input to mid-voltage (which is earth voltage when balanced power supplies are used) and using 100% d.c. feedback from the output to the (−) input. The gain is given by:

$$G = R_1/R_2$$

Note that a capacitor C_1 is needed when a single-ended power supply is used to prevent the d.c. bias voltage from being divided down in the same ratio as the a.c. bias. When balanced power supplies are used, direct coupling is possible provided that the signal source is at zero d.c. volts.

The input resistance for these circuits is simply the value of resistor R_2, since the effect of the feedback is to make the input resistance at the (−) input almost zero; this point is referred to as a 'virtual earth' for signals. The output resistance is typically about 150 ohms.

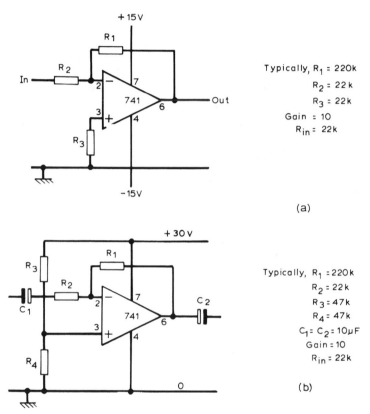

Typically, $R_1 = 220k$
$R_2 = 22k$
$R_3 = 22k$
Gain $= 10$
$R_{in} = 22k$

(a)

Typically, $R_1 = 220k$
$R_2 = 22k$
$R_3 = 47k$
$R_4 = 47k$
$C_1 = C_2 = 10\mu F$
Gain $= 10$
$R_{in} = 22k$

(b)

Figure 4.3. Inverting amplifier configuration. (a) uses balanced power supplies. Ideally, R_3 should equal R_2, though differing values are often used. The gain is set by the ratio R_1/R_2, and the input resistance is equal to R_2. Using an unbalanced power supply (b), the + input is biased to half the supply voltage (15 V in this example) by using equal values for R_3 and R_4. The gain is again given by R_1/R_2. Coupling capacitors are needed because of the d.c. bias conditions

Circuits for non-inverting amplifiers are shown in *Figure 4.4*. Non-inverting amplifiers also make use of negative feedback to stabilise the working conditions in the same way as the inverting amplifier circuits, but the signal input is now to the (+) input terminal. The gain is

$$G = \frac{R_1 + R_2}{R_2}$$

and the circuit is sometimes referred to as the 'voltage-follower with gain'. The input resistance is high, usually around 1 MΩ, for the dual-supply version, though the bias resistors (*Figure 4.4b*) reduce this to a few hundred kΩ.

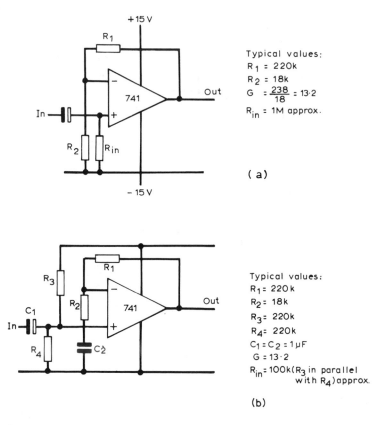

Typical values:
$R_1 = 220k$
$R_2 = 18k$
$G = \dfrac{238}{18} = 13\cdot2$
$R_{in} = 1M$ approx.

(a)

Typical values:
$R_1 = 220k$
$R_2 = 18k$
$R_3 = 220k$
$R_4 = 220k$
$C_1 = C_2 = 1\mu F$
$G = 13\cdot2$
$R_{in} = 100k (R_3$ in parallel with R_4) approx.

(b)

Figure 4.4. *Non-inverting amplifiers. Using a balanced power supply (a), only two resistors are needed, and the voltage gain is given by* $\dfrac{R_1 + R_2}{R_2}$. *The input resistance is very high. When an unbalanced supply (b) is used, a capacitor C_2 must be connected between R_2 and earth to ensure correct feedback of signal without disturbing bias. The input resistance is now lower because of R_3 and R_4 which as far as signal voltage is concerned are in parallel*

Figure 4.5 shows the 741 used as a differential amplifier, though with a single-ended output. The gain is set by the ratio R_1/R_2 as before — note the use of identical resistors in the input circuits to preserve balance.

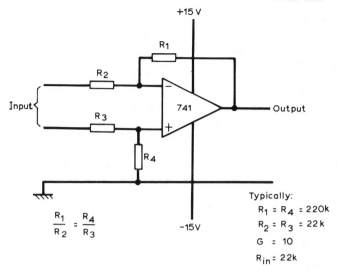

Figure 4.5. Differential amplifier application. Both inputs are used for signals which must be in antiphase (balanced about earth). Any common-mode signals (in phase at both inputs) are greatly attenuated

General notes on op-amp circuits

The formulae for voltage gain hold for values of gain up to several hundred times, because the gain of the op-amp used *open-loop* (without feedback) is very high, of the order of 100 000 (100 dB). The maximum load current is about 10 mA, and the maximum power dissipation 400 mW. The 741 circuit is protected against damage from short-circuits at the output, and the protection circuits will operate for as long as the short-circuit is maintained.

The frequency range of an op-amp depends on two factors, the gain-bandwidth product for small signals, and the slew rate for large signals. The gain-bandwidth product is the quantity, $G \times B$, with G equal to voltage gain (not in dB) and B the bandwidth upper limit in Hz. For the 741, the $G-B$ factor is typically 1 MHz, so that, in theory, a bandwidth of 1 MHz can be obtained when the voltage gain is unity, a bandwidth of 100 kHz can be attained at a gain of ten, a bandwidth of 10 kHz at a gain of 100 times, and so on. This trade-off is usable only for small signals, and cannot necessarily be applied to all types of operational amplifiers. Large amplitude signals are further limited by the slew rate of the circuits within the amplifier. The slew rate of an amplifier is the value of maximum rate of change of output voltage

$$\text{slew rate} = \frac{\text{maximum voltage change}}{\text{time needed}}$$

Units are usually volts per microsecond.

Because this rate cannot be exceeded, and feedback has no effect on slew rate, the bandwidth of the op-amp for large signals, sometimes called the power bandwidth, is less than that for small signals. The slew rate limitation cannot be corrected by the use of negative feedback; in fact negative feedback acts to increase distortion when the slew rate limiting action starts, because the effect of the feedback is to increase the rate of change of voltage at the input of the amplifier whenever the rate is limited at the output. This accelerates the overloading of the amplifier, and can change what might be a temporary distortion into a longer-lasting overload condition.

The relationship between the sine wave bandwidth and the slew rate, for many types of operational amplifier is:

$$\text{Max. slew rate} = 2\,\pi\,f_{max}\,E_{peak}$$

or

$$f_{max} = \frac{\text{max. slew rate}}{2\,\pi\,E_{peak}}$$

where slew rate is in units of volts per second (*not* V/μs), f_{max} is the maximum full-power frequency in Hz, and E_{peak} is the peak voltage of the output sine wave.

This can be modified to use slew rate figures in the more usual units of V/μs, with the answer in MHz. For example, a slew rate of 1.5 V/μs corresponds to a maximum sine wave frequency (at 10 V output) of

$$f_{max} = \frac{1.5}{2\,\pi \times 10}\ \text{MHz} = 0.023\ \text{MHz or 23 kHz.}$$

Slew rate limiting arises because of internal stray capacitances which must be charged and discharged by the current flowing in the transistors inside the i.c.: improvement is obtainable only by redesigning the internal circuitry. The 741 has a slew rate of about 0.5 V/μs, corresponding to a power bandwidth for 12 V peak sine wave signals of about 6.6 kHz. The slew rate limitation makes op-amps unsuitable for applications which require fast-rising pulses, so that a 741 should not be used as a signal source or feed (interface) with digital circuitry, particularly TTL circuitry, unless a Schmitt trigger stage is also used.

Higher slew rates are obtainable with more modern designs of op-amps; for example, the Fairchild LS201 achieves a slew rate of 10 V/μs.

Other operational amplifier circuits

Figures 4.6 to *4.9* illustrate circuits other than the straightforward voltage amplifier types. *Figure 4.6* shows two versions of follower circuit with no voltage gain, but with useful characteristics. The unity gain inverter will provide an inverted output of exactly the same amplitudes as the input signal, subject to slew rate limitations. The non-inverting circuit, or voltage follower, performs the same action as the familiar emitter-follower, having a very high input resistance and a low output resistance. For this type of circuit, the action of the feedback causes both inputs to change voltage together, as a common-mode signal

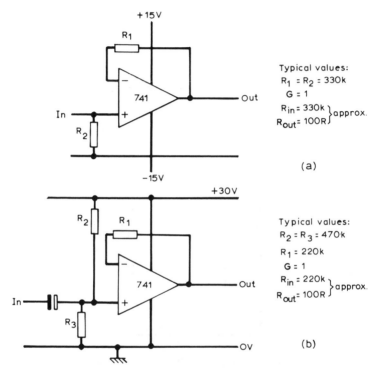

Figure 4.6. The voltage follower. The gain is unity, with high input resistance and low output resistance. The dual-voltage supply version (a) uses only two resistors whose values are not critical. Ideally, R_1 should equal R_2, and both should be high values so that the input resistance is high, equal to R_2. The single supply voltage version (b) uses three resistors, with $R_2 = R_3$ and R_1 made equal to $R_2/2$, which is also the value of input resistance

would, so that any restrictions on the amplitude of common-mode signals (see the manufacturer's sheets) will apply to this circuit. *Figure 4.7* shows two examples of a 741 as it is used in a variety of 'shaping' circuits in which the gain/frequency or gain/amplitude graph is intended to be non-linear.

The use of op-amps for switching circuits is limited by the slew rate, but the circuits shown in *Figures 4.8* and *4.9* are useful if fast-rising or falling waveforms are not needed.

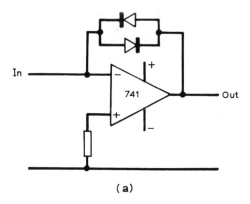

(a)

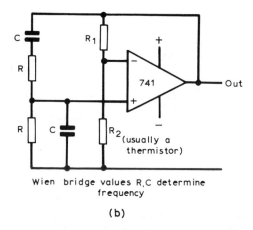

Wien bridge values R,C determine
frequency

(b)

Figure 4.7. Using the 741 in circuits which are not linear
amplifiers. (a) A constant-output amplifier. Because the
diodes will permit feedback of voltages whose amplitude is
enough to allow the diodes to conduct, the output voltage is
limited to about this amplitude, but without excessive
clipping. The gain is very large for small input signals, and
very small for large input signals. (b) The Wien bridge in the
feedback network causes oscillation. The waveform is a sine
wave only if the gain is carefully controlled by making
$R_1/R_2 = 3$, and this is usually done by making R_2 a
thermistor whose resistance value decreases as the voltage
across it increases. The frequency of oscillation is given by

$$f = \frac{1}{2 \pi RC}$$

Current differencing amplifiers

A variation on the op-amp circuit uses current rather than voltage input signals, and is typified by the National Semiconductor LM3900. In this i.c., which contains four identical op-amps, the + and − inputs are current inputs, whose voltage is generally about +0.6 V when correctly biased. A single-ended power supply is used, and the output voltage can reach to within a fraction of a volt of the supply limits. The output voltage is proportional to the difference between the *currents* at the

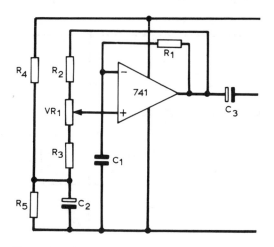

Figure 4.8. A 741 used as an astable multivibrator; a single power supply voltage version is illustrated. The use of R_2, VR_1, R_3 in the positive feedback path sets the + input to a definite fraction of the output voltage above or below the normal half-supply-voltage. When the output changes over, C_1 charges through R_1 until the − input reaches the same voltage as the + input upon which the circuit switches over. The voltage change at the + input is then rapid, but the voltage at the − input cannot change until C_1 has charged again. The frequency is therefore set by the time constant C_1R_1 and also by the setting of VR_1

two inputs, so that bias conditions are set by large value resistors. *Figure 4.10* shows a typical amplifier circuit, in which the current into the (+) input is set by R_1, whose value is 2.2 MΩ. Because the ideal bias voltage for the output is half of supply voltage, a 1 MΩ resistor is used connected between the output and the (−) input. In this way, the currents to the two inputs are identical, and the amplifier is correctly biased. The advantages of this type of op-amp are now being recognised, and an equivalent to the LM3900 is now available from Motorola.

Other linear amplifier i.c.s

A very large variety of i.c.s intended for a.f., i.f. and r.f. amplifiers can be obtained. For any design work, the full manufacturer's data sheet pack must be consulted, but a few general notes can be given here. A.F. i.c. circuits use direct coupling internally, because of the difficulty of fabricating capacitors of large value on to silicon chips, but the high gains which are typical of operational amplifiers are not necessary for most a.f. applications. Faster slew rates and greater open-loop band-widths can therefore be attained than is practicable using op-amps.

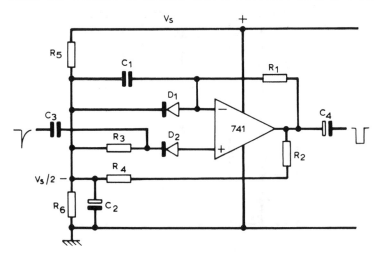

Figure 4.9. A 741 monostable. With no input, the output voltage is high, which causes the + input voltage to be higher than the voltage, $V_s/2$ set by R_5 and R_6 (equal values). Because of D_1, the − input voltage cannot rise to the same value as the + input. A negative pulse at the + input causes the output voltage to drop rapidly, taking the + input voltage low. The − input voltage then drops at a rate determined by the time constant C_1R_1. When the − input voltage equals the + input voltage, the circuit switches back, and the diode D_1 conducts to 'catch' the − input voltage and so prevent continuous oscillation

Many a.f. i.c.s use separate chips for preamplifier and for power amplifier uses, with separate feedback loops for each. Frequency correcting networks composed of resistors and capacitors are usually needed to avoid oscillation, and heatsinks will be needed for the larger power amplifier i.c.s. The need for external volume, stereo balance, bass and treble controls, along with feedback networks, makes the circuitry rather more involved than some other i.c. applications.

Figure 4.11 shows two a.f. circuits examples. Note that the stability of these audio i.c.s is often critical, and decoupling capacitors, as specified by the manufacturers, must be connected as close to the i.c.

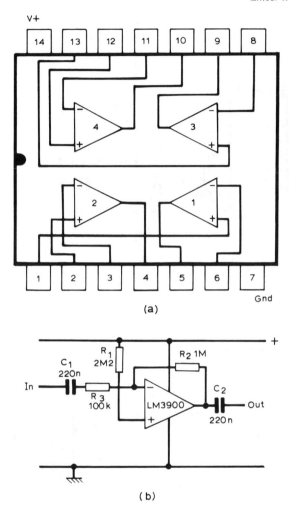

Figure 4.10. The current differencing amplifier, or Norton Op-amp. (a) Pinout for the LM3900, which contains four amplifiers in a single fourteen-pin package. (b) Typical amplifier circuit. Note the high resistor values

pins as possible. For stability reasons, also, stripboard construction is extremely difficult with some i.c. types, and suitable printed-circuit boards should be used.

I.F. and r.f. amplifier circuits contain untuned wideband amplifier circuits to which tuning networks, which may be *LC* circuits or trans-filters, may be added. It is possible to incorporate r.f., mixer, i.f. and

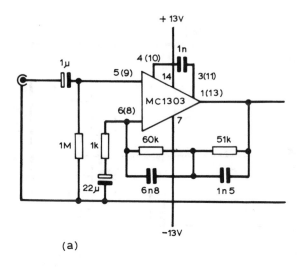

(a)

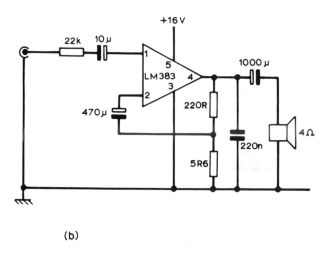

(b)

Figure 4.11. Audio amplifier i.c.s. (a) The MC1303 preamplifier is a dual unit for stereo use — the pin numbers in brackets are for the second section. Inputs up to 5 mV can be accepted, and the circuit here is shown equalised for a magnetic pickup. The output is 250 mV with a 5 mV input at a distortion level of about 0.1%. (b) The LM383 power amplifier uses a five-pin TO 220 package. The power output is 7 W into 4 ohms, with a distortion level of 0.2% at 4 W output. The maximum power dissipation is 15 W when a 4 °C/W heatsink is used

demodulator stages into a single i.c., but generally only when low-frequency r.f. and i.f. are used. A very common scheme for f.m. radio receivers is to use a discrete component tuner along with i.c. i.f. and demodulator stages, using the usual 10.7 MHz i.f. *Figure 4.12* shows an example of such an i.f. stage. Once again, when a large amount of gain

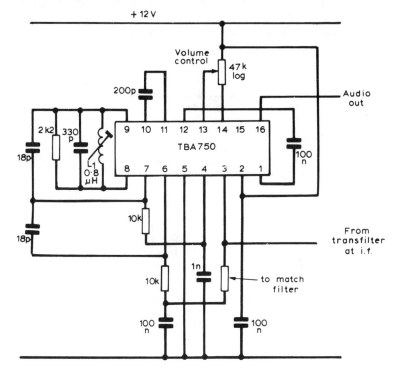

Figure 4.12. An i.f. detector i.c. for use in 10.7 MHz stereo f.m. i.f. stages. The minimum input for limiting is 100 μV, and the volume control range (operating on d.c.) is 80 dB. The audio output is 1.4 V r.m.s. with a signal of 15 kHz deviation

is attained in one i.c. stability is a major problem, and the manufacturer's advice on decoupling must be carefully followed. At the higher frequencies, the physical layout of components is particularly important, so that p.c.b.s intended for the TBA750 i.c. (and similar) should be used rather than stripboards.

Phase locked loops

The phase-locked loop is a type of linear i.c. which is now being used to a considerable extent. The block diagram of the circuit is outlined in

Figure 4.13 and consists of a voltage-controlled oscillator, a phase-sensitive detector, and comparator units. The oscillator is controlled by external components, so that the frequency of oscillation can be set by a suitable choice of these added components. An input signal to the Phase Locked Loop (PLL) is compared in the phase-sensitive detector to the frequency generated in the internal oscillator, and a voltage output obtained from the phase-sensitive detector. Provided that the input frequency is not too different from the internally generated

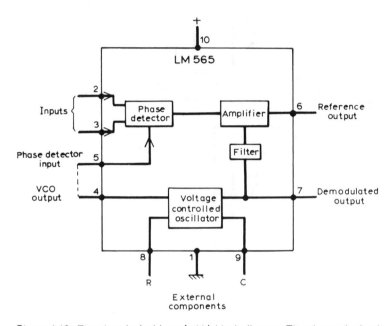

Figure 4.13. The phase-locked loop (p.l.l.) block diagram. The pin numbering is for the LM565. The signal input can be to pin 2 or 3 in this i.c., and in normal use pins 4 and 5 are linked

frequency (within the 'pull-in' range), the voltage from the phase-sensitive detector can then be used to correct the oscillator frequency until the two signals are at the same frequency and in the same phase. Either the oscillator or the correcting voltage may be used as an output. The circuit can be used, for example, to remove any traces of amplitude modulation from an input signal, since the output (from the internal oscillator) is not affected by the amplitude of the input signal, but is locked to its frequency and phase. The circuit may also be used as an f.m. demodulator, since the control voltage will follow the modulation of an f.m. input in its efforts to keep the oscillator locked in phase.

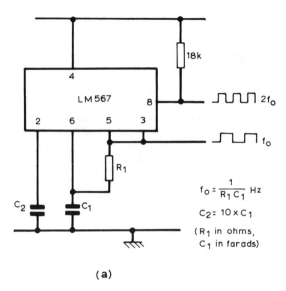

(a)

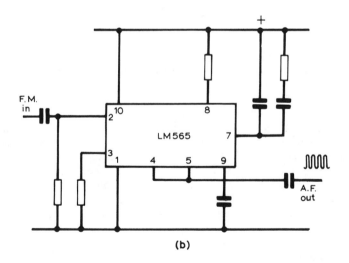

(b)

Figure 4.14. P.L.L. circuits. (a) Oscillator with fundamental and second harmonic outputs. (b) F.M. demodulator — the component values must be calculated with reference to the i.f. frequency which is used — this cannot be as high as the normal 10.7 MHz because the operating frequency limit of this i.c. is 500 kHz

Voltage stabiliser i.c s

The ease with which zener diode junctions and balanced amplifiers may be constructed in integrated form, together with the increasing demand for stabilised supplies and the steady increase in the power which can be dissipated from i.c.s due to improved heat-sinking methods, has led to the extensive use of i.c. voltage stabiliser circuits, to the extent that discrete component stabilisers are almost extinct. The decisive factor is that i.c.s can include such features as overload and short-circuit protection at virtually no extra cost, using an elaborate circuit such as that of the SGS-ATES TBA625A shown in *Figure 4.15*. The overload

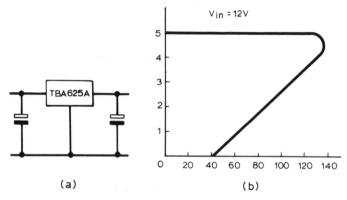

Figure 4.15. Voltage regulation. (a) Use of an i.c. regulator, the TBA625A, is illustrated. (b) Foldback overcurrent protection — at excessive currents the voltage output and current output both drop as indicated in the graph

protection is usually of the 'foldback' type, illustrated by the characteristic of *Figure 4.15b*, in which excessive current causes the output voltage to drop to zero with a much smaller current flowing. *Figure 4.16* shows circuit applications — note that a fixed voltage regulator can be used to provide an adjustable output, and higher current operation can be achieved by adding power transistors to the circuit. Voltage stabiliser i.c.s are available for all the commonly used voltage levels.

Motor-speed controllers are a more specialised form of stabiliser circuit, and are used to regulate the speed of d.c. motors in record players, tape and cassette recorders, and model motors. A typical application is shown in *Figure 4.17*.

One very important class of linear i.c.s is concerned with television circuitry. The development of linear i.c.s has been such that virtually every part of a TV circuit with the exceptions of the tuner head and the horizontal output stage can now be obtained in i.c. form. Because of the specialised nature of such circuits, the reader is referred to the

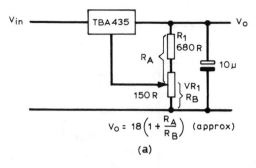

$$V_O = 18\left(1 + \frac{R_A}{R_B}\right) \quad \text{(approx)}$$

(a)

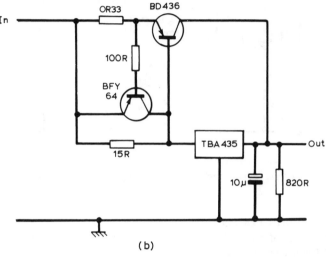

(b)

Figure 4.16. Extending the range of fixed-voltage regulators. (a) Extending voltage range, (b) extending current range

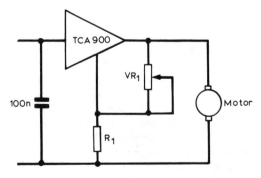

Figure 4.17. A motor-speed control i.c. The values of VR_1 and R_1 must be calculated with reference to the resistance of the motor windings

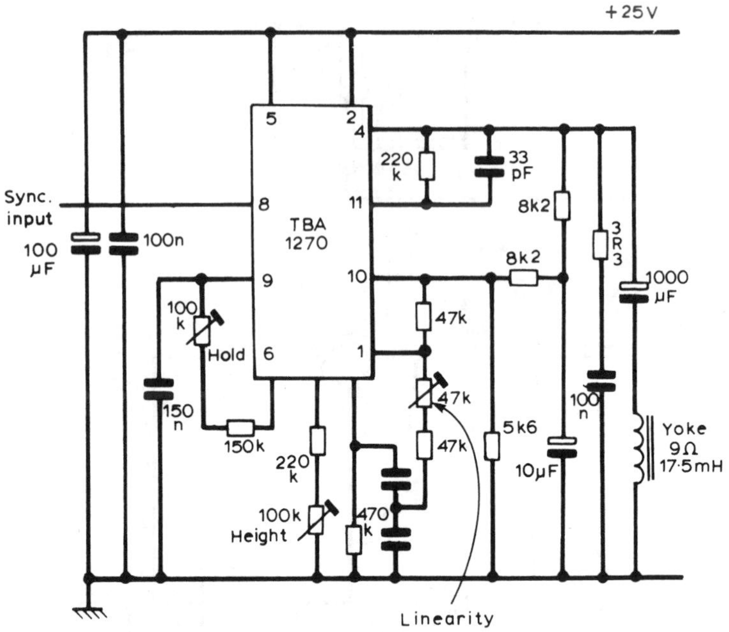

Figure 4.18. The TDA 1270 used for vertical deflection in a 12 in portable TV receiver

manufacturer's handbooks for further information, but as an example of the uses of such i.c.s, *Figure 4.18* shows a TDA1270 (SGS-ATES) which performs the functions of vertical oscillator, timebase generator and output stage.

The 555 timer

This circuit is generally classed among linear circuits because it uses op-amp circuits as comparators. The purpose of the timer is to generate time delays or waveforms which are very well stabilised against voltage changes. A block diagram of the internal circuits is shown in *Figure 4.19*.

A negative-going pulse at the trigger input, pin 2, makes the output of comparator (2) go high. The internal resistor chain holds the (+) input of comparator (2) at one third of the supply voltage, and the (−) input of comparator (1) at two-thirds of supply voltage, unless pin 5 is connected to some different voltage level. The changeover of comparator (2) causes the flip-flop to cut off Tr_1, and also switch the output stage to its high-voltage output state. With Tr_1 cut off, the external capacitor C can charge through R (also external) until the voltage at pin

6 is high enough, equal to 2/3 of the supply voltage, to operate comparator (1). This resets the flip-flop, allows Tr_1 to conduct again, so discharging C, and restores the output to its low voltage state. Resetting is possible during the timing period by applying a negative pulse to the reset pin, number 4.

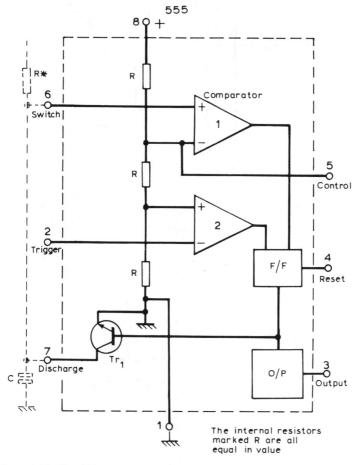

Figure 4.19. The 555 timer block diagram. R and C are external components which are added in most applications of the timer*

The triggering is very sensitive, and some care has to be taken to avoid unwanted triggering pulses, particularly when inductive loads are driven. Retriggering caused by the back-e.m.f. pulse from an inductive load is termed 'latch-up', and can be prevented by the diode circuitry shown in *Figure 4.20*.

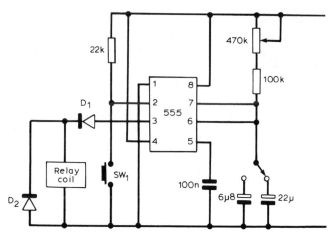

D₂ should be a gold-bonded germanium diode

Figure 4.20. A relay-timer circuit using the 555. On pressing the switch, the relay is activated for a time determined by the setting of the 470 kΩ variable and the capacitor value selected by the switch. Note the use of diodes to prevent latch-up and damage to the i.c. when the relay is switched off

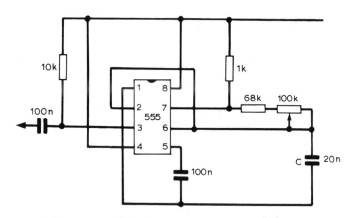

Figure 4.21. An astable pulse generator, with variable frequency output controlled by the 100 kΩ potentiometer. The capacitor C can be a switched value if desired

Two typical circuits are shown in *Figures 4.21* and *4.22*. The timer is available from several manufacturers, all using the same 555 number though prefixed with different letter combinations which indicate the manufacturer.

Chapter 5

Digital I.C.s

Basic logic notes

Unlike linear circuits, digital circuits process signals which consist of only two significant voltage levels, labelled logic 0 and logic 1. Most logic systems use positive logic, in which logic 0 is represented by zero volts, or a low voltage, below 0.5 V; and logic 1 is represented by a higher voltage. Changes of level from 0 to 1 or 1 to 0 are made as quickly as possible, since slow changes can cause faulty operation of some types of logic circuits.

The use of two logic levels naturally leads to the use of a scale-of-two or binary scale for counting. In a binary scale, the only digits used are 1 and 0, with the placing of the 1 indicating what power of 2 is represented. *Table 5.1* shows methods of converting to and from binary numbers and decimal numbers.

In addition, because large binary numbers are awkward to handle, and difficult to copy without error, hexadecimal (or hex) numbers are used for many applications, particularly in microprocessor machine code (see later). Hex coding is used when binary numbers occur in groups of four, eight (called a byte) or multiples, and the conversions are shown in *Table 5.2*. The use of hex coding makes the tabulation of binary numbers considerably simpler, but the circuits to which the hex codes refer will still operate in binary.

Note in this context that a three-state or tri-state device does not imply a third logic voltage. This description refers to a circuit which can be switched to a high impedance at an input or output (or both) so that it does not affect or respond to the voltages of signals connected to it.

Table 5.1. CODE CONVERSION

DECIMAL-TO-BINARY CONVERSION

Write down the decimal number. Divide by two, and write the result underneath, with the remainder, 0 or 1, at the side. Now divide by two again, placing the new (whole number) result underneath, and the remainder, 0 or 1, at the side. Repeat until the last figure (which will be 2 or 1) has been divided, leaving zero.

Now read the remainders in order from the foot of the column upwards to give the binary number.

Example: Convert decimal 1065 into binary:

1065	
532	1
266	0
133	0
66	1
33	0
16	1
8	0
4	0
2	0
1	0
0	1

Binary number is 10000101001

BINARY-TO-DECIMAL CONVERSION

Using the table of binary powers, which shows the values of successive powers of two, starting from the right-hand side of the binary number (the least significant figure). Write down the decimal equivalent for each 1 in the binary number, and then add.

Example: 1 0 0 0 0 1 0 1 0 0 1

	Decimal
	1
	8
	32
	1024
Total	1065

Power of two	Decimal	Power of two	Decimal
0	1	10	1 024
1	2	11	2 048
2	4	12	4 096
3	8	13	8 192
4	16	14	16 384
5	32	15	32 768
6	64	16	65 536
7	128	17	131 072
8	256	18	262 144
9	512	19	524 288

Table 5.2. BINARY, HEXADECIMAL AND DECIMAL

4-bit Binary	Decimal	Hexadecimal
0000	0	0
0001	1	1
0010	2	2
0011	3	3
0100	4	4
0101	5	5
0100	6	6
0111	7	7
1000	8	8
1001	9	9
1010	10	A
1011	11	B
1100	12	C
1101	13	D
1110	14	E
1111	15	F

To convert binary to hex: arrange the binary numbers in four-bit groups starting from the right-hand side (the least significant digit). Convert each group to hex, using the table above.

Examples: 01011101 converts to 5D,   11100011 converts to E3

To convert hex to binary: write down the equivalent for each number, using the table above.

Examples: E1 = 11100001,   6F = 01101111

Combinational logic circuits are those in which the output at any time is determined entirely by the combination of input signals which are present at that time. A combinational logic circuit can be made to produce a logic 1 at its output only for some set pattern of 1s and 0s at various inputs, just as a combination lock will open only when the correct pattern of numbers has been set. One of the major uses for

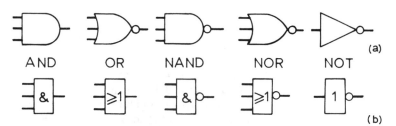

Figure 5.1. Gate symbols. (a) International (U.S. MIL) symbols, used in magazine articles, most books and manufacturer's manuals, (b) British Standard (B.S.) symbols used for TEC and C & G examinations and textbooks

combinational logic circuits is in recognising patterns, whether these be binary numbers or other information in binary form.

Any required combinational logic circuit can be built up from a few basic circuits, the choice of which is dictated more by practical than by theoretical considerations. The three most basic circuits are those of the

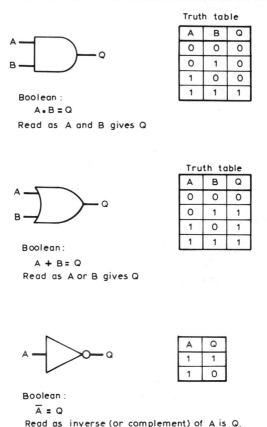

Truth table

A	B	Q
O	O	O
O	1	O
1	O	O
1	1	1

Boolean :
A . B = Q
Read as A and B gives Q

Truth table

A	B	Q
O	O	O
O	1	1
1	O	1
1	1	1

Boolean:
A + B = Q
Read as A or B gives Q

A	Q
1	1
1	O

Boolean :
$\overline{A} = Q$
Read as inverse (or complement) of A is Q,
or as   Q = NOT A

Figure 5.2. Truth tables and Boolean expressions for three basic gates

AND, OR and NOT gates; the NOT gate is also referred to as an inverter or complementer. These are represented in circuits by the symbols shown in *Figure 5.1*; the international symbols are much more commonly used. The internal circuitry of the i.c.s is not usually shown, since the action of the circuits is standardised. All that need be known about the internal circuits is the correct level of power supplies, driving signals and output signals, together with any handling precautions.

The action of a logic gate or circuit can be described by a truth table or by a Boolean expression. A truth table shows each possible input to the logic circuit with the output which such a set of inputs produces; examples are shown in *Figure 5.2*. A Boolean expression is a much shorter way of showing the action, using the symbols + to mean OR and . to mean AND. The action of a four-input AND gate, for example, can be written as A.B.C.D = 1; a truth table for this gate would take up half a page, because the number of lines of truth table is equal to 2^n, where *n* is the number of inputs.

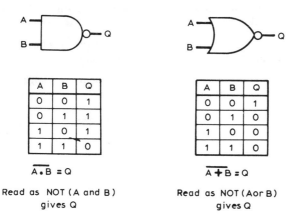

A	B	Q
0	0	1
0	1	1
1	0	1
1	1	0

$\overline{A.B} = Q$

Read as NOT (A and B)
gives Q

A	B	Q
0	0	1
0	1	0
1	0	0
1	1	0

$\overline{A+B} = Q$

Read as NOT (A or B)
gives Q

Figure 5.3. NAND and NOR gate truth tables and Boolean expressions

The NAND and NOR gates shown in *Figure 5.3* combine the actions of the AND and OR gates with that of an inverter. These gates are simpler to produce, and either can be used as an inverter so that for practical purposes these are more common than the ordinary AND and OR gates. Another type of circuit which is used to such an extent that it is available in i.c. form is the exclusive-OR (X-OR) gate. The truth table for a two-input X-OR gate is shown in *Figure 5.4*, in comparison

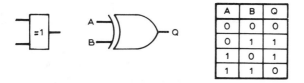

A	B	Q
0	0	0
0	1	1
1	0	1
1	1	0

Boolean : $(\overline{A.B}).(A+B) = Q$

Figure 5.4. Exclusive-OR (X-OR) gate. Symbols, truth table and Boolean expression

with that for a simple OR-gate. The Boolean expression is $Q = (A + B)$. $\overline{(A.B)}$, where the bar above the brackets indicates NOT (= inverse). The expression A.B will be 1 when A = 1 and B = 1, and the inverse bar indicates that the output must be zero for this state. This quantity is ANDED with (A or B), so as to make the output zero for A = 1 and B = 1.

The advantages of setting out logic circuit requirements as Boolean expressions rather than in truth tables are:

(1)   The Boolean expressions are considerably quicker to write.

(2)   The Boolean expressions can be simplified (often) by applying a set of rules.

(3)   The Boolean expressions can usually indicate what combination of single gates will be needed for the circuit.

The rules of Boolean algebra (invented, incidentally, long before digital logic circuits existed) are shown in *Table 5.3*. The usefulness of these

Table 5.3. BOOLEAN ALGEBRA

1. Definitions of gate action

$0 + 0 = 0$	$0.0 = 0$
$0 + 1 = 1$	$0.1 = 0$
$1 + 0 = 1$	$1.0 = 0$
$1 + 1 = 1$	$1.1 = 1$

2. For gates to which one input is a variable A (which can be 0 or 1)

$A + 0 = A$	$A.0 = 0$
$A + 1 = 1$	$A.1 = A$
$A + A = A$	$A.A = A$
$A + \overline{A} = 1$	$A.\overline{A} = 0$
$\overline{\overline{A}} = A$	

3. For more than one variable (each of which can be 0 or 1)

$A + B = B + A$	$A.B = B.A$
$(A + B) + C = A + (B + C)$	$(A.B).C = A.(B.C)$
$(A.B) + (A.C) = A.(B + C)$	$(A + B).(A + C) = A + (B.C)$

4. De Morgan's theorem

$\overline{A.B} = \overline{A} + \overline{B}$	$\overline{A + B} = \overline{A}.\overline{B}$

rules is that they may be applied to simplify an expression, so saving considerably in the number of logic gates that are needed to carry out a logic operation. For example, the proposed gate system in *Figure 5.5* carries out the process which in Boolean algebra becomes

$$(A . B . C) + A . (\overline{B} + \overline{C})$$

Leaving the first term unchanged, we can apply Rule 4 to the second term

$(A . B . C) + A . (\overline{B . C})$

Now taking out the common factor A changes the expression to

$A . (B . C + \overline{B . C})$ (Rule 3)

$= A . 1$ (Rule 2)

$= A$ (Rule 2)

From this, it appears that only the A signal is needed. An experienced designer might, in fact, be able to deduce that the B and C signals were

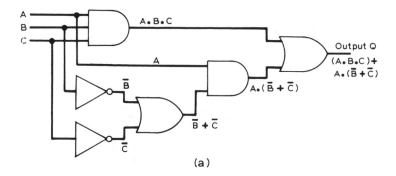

(a)

A	B	C	A·B·C	$\overline{B}$	$\overline{C}$	$\overline{B}+\overline{C}$	$A·(\overline{B}+\overline{C})$	Q
0	0	0	0	1	1	1	0	0
0	0	1	0	1	0	1	0	0
0	1	0	0	0	1	1	0	0
0	1	1	0	0	0	0	0	0
1	0	0	0	1	1	1	1	1
1	0	1	0	1	0	1	1	1
1	1	0	0	0	1	1	1	1
1	1	1	1	0	0	0	0	1

(b)

Figure 5.5. Analysing the action of a gate system. (a) The text shows the method using Boolean algebra, the truth table shows that the Q output column corresponds exactly with the A input column. The truth table method is simple, but very tedious, because the number of lines of truth table are equal to 2^n, where n is the number of inputs (A, B, C, etc.)

redundant by an inspection of the gate circuit, but the Boolean algebra is often faster to write, if nothing else!

By way of another example, consider the gates of *Figure 5.6*, which implement the Boolean expression

$Q = \overline{A} \cdot C \cdot + \overline{A} \cdot B \cdot D + A \cdot C \cdot D + A \cdot B \cdot D$

Taking out the common factor D, this becomes

$$Q = D \cdot (\overline{A} \cdot C + \overline{A} \cdot B + A \cdot C + A \cdot B)$$

and taking out the next common factor (C + B), this transforms to

$$Q = D \cdot (\overline{A} \cdot (C + B) + A \cdot (C + B))$$

which is

$$Q = D \cdot (C + B) \cdot (\overline{A} + A)$$

$$= D \cdot (C + B), \text{ since } \overline{A} \text{ or } A \text{ must be logic 1.}$$

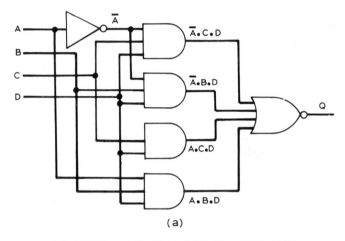

(a)

A	B	C	D	$\overline{A}$	$\overline{A} \cdot C \cdot D$	$\overline{A} \cdot B \cdot D$	$A \cdot C \cdot D$	$A \cdot B \cdot D$	Q
0	0	0	0	1	0	0	0	0	0
0	0	0	1	1	0	0	0	0	0
0	0	1	0	1	0	0	0	0	0
0	0	1	1	1	1	0	0	0	1
0	1	0	0	1	0	0	0	0	0
0	1	0	1	1	0	1	0	0	1
0	1	1	0	1	0	0	0	0	0
0	1	1	1	1	1	1	0	0	1
1	0	0	0	0	0	0	0	0	0
1	0	0	1	0	0	0	0	0	0
1	0	1	0	0	0	0	0	0	0
1	0	1	1	0	0	0	1	0	1
1	1	0	0	0	0	0	0	0	0
1	1	0	1	0	0	0	0	1	1
1	1	1	0	0	0	0	0	0	0
1	1	1	1	0	0	0	1	1	1

(b)

Figure 5.6. A gate system (a) whose action is analysed by Boolean algebra in the text. The truth table (b) shows that the output is logic 1 when D is 1 and C or B also 1

For really large and complex circuits, Boolean algebra can be tedious and difficult, but other approaches are equally tedious and difficult — there is no easy way of dealing with complex circuits.

Note that in circuits which use an elaborate sequence of gates, problems can arise because of the small but significant time delay which

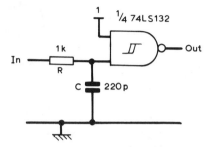

Figure 5.7. A circuit for suppressing unwanted short pulses. This also inverts, so that a second Schmitt inverter may be needed after the Schmitt gate

is introduced by each gate. These 'race hazard' problems arise when the final gate of a set is switched by signals which have passed through different numbers of gates in their different paths. A time delay between these paths can cause a brief unwanted pulse, of duration equal to the time delay between the input paths, at the output. This unwanted pulse may cause problems, particularly in a counting circuit which follows the gate circuits. Methods of eliminating race hazards are too complex to discuss in this book, but *Figure 5.7* shows a simple unwanted-pulse suppressor which can be used for comparatively slow logic systems. The output is integrated by *R*, *C*, using a time constant chosen so that a short pulse (50 ns or so) will have little effect, but the delay which the circuit causes to a longer pulse is not serious. The sharp rise and fall times of the logic signals then have to be restored by using a Schmitt NAND gate, IC1.

Sequential logic

Sequential logic circuits change output when the correct sequence of signals appears at the inputs. The simplest of the sequential logic circuits is the R-S (or S-R) flip-flop, or latch, circuit shown in symbol form in *Figure 5.8a* with one possible circuit using NAND gates. The truth table for this device is shown in *Figure 5.8b*, from which it can be seen that it is the sequence of signals at the two inputs rather than the signal levels themselves which decide the output. Note that the outputs Q and $\overline{Q}$ are intended to be complementary (one must be the inverse

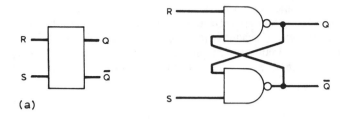

(a)

Truth table

R	S	Q	Q̄
O	O	1✳	1✳
O	1	1	O
1	1	1	O
1	O	O	1
1	1	O	1

✳Undesirable state

The R=0, S=0 state must be avoided because we normally want Q̄ to be the inverse of Q

(b)

Figure 5.8. The R-S flip-flop. (a) Symbol and one possible circuit (NOR gates can also be used). (b) Truth table. The output for R=1, S=1 depends on the previous values of R and S

of the other), so that the state R = 0, S = 0 must never be allowed, since it results in Q = 1, Q̄ = 1.

The R-S flip-flop is used to lock or latch information, one bit to each R-S latch, because the output of the latch is held fixed when both

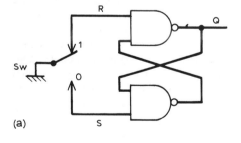

(a)

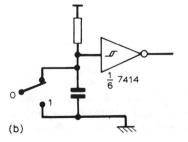

(b)

Figure 5.9. A 'switch-debounce' circuit using an R-S flip-flop (a). An alternative arrangement which needs only an on/off switch type is shown in (b) — this uses a Schmitt inverter rather than the R-S flip-flop

inputs are at logic 1 in the state which existed just before the second input went high. One typical application is to suppress the effects of contact bounce in mechanical switches, as shown in *Figure 5.9*. The effect of bounce in this arrangement is to raise both inputs of the latch to logic 1, leaving the outputs unaffected.

The applications of the simple R-S latch are rather limited, and most sequential logic circuits make use of the principle of clocking. A clocked circuit has a terminal, the clock (CK) input, to which

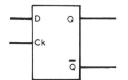

D	Q_n	Q_n+1
0	0	0
0	1	0
1	0	1
1	1	1

Figure 5.10. Symbol for D-type flip-flop and truth table. In the truth table, Q_n denotes the output before the clock pulse, and Q_n+1 the output after the clock pulse

rectangular pulses, the clock pulses) can be applied. The circuit actions, which is determined by the other inputs, takes place only in the time of the clock pulse, and may be synchronised to the leading edge, the trailing edge, or the time when the clock pulse is at logic 1, according to the design of the circuit.

The D-type flip-flop, shown in *Figure 5.10*, with its truth table, is one form of D-type clocked flip-flop. The output at Q will become identical to the input at D (for DATA) only at the clock pulse, generally at the leading edge of the pulse, so earning the circuit the name 'edge-triggered'. By connecting the $\bar{Q}$ output back to the D-input, the flip-flop will act as a bistable counter, or divide-by-two circuit, because the voltage at Q will cause a change of state only when the leading edge of the clock-pulse occurs. The $\bar{Q}$ voltage itself will change only a little time after the leading edge of the clock pulse, so that there is no effect on the flip-flop. A few D-type flip-flops change state at the trailing edge of the clock pulse, and some allow the D-input to affect the Q-output for as long as the clock pulse is at logic 1. This latter type is sometimes called a 'transparent latch'.

The J-K flip-flop (*Figure 5.11a*) is a much more flexible design which uses a clock pulse along with two control inputs labelled J and K. The internal circuit makes use of two sets of flip-flops, designated master and slave respectively. At the leading edge of the clock pulse, the master flip-flop changes state under the control of the inputs at J and K. After this time, changes at J and K have no effect on the master flip-flop. There is no change at the output, however, until the trailing

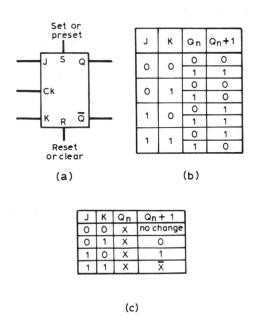

(a)

J	K	Q_n	Q_n+1
0	0	0	0
		1	1
0	1	0	0
		1	0
1	0	0	1
		1	1
1	1	0	1
		1	0

(b)

J	K	Q_n	Q_n+1
0	0	X	no change
0	1	X	0
1	0	X	1
1	1	X	$\overline{X}$

(c)

Figure 5.11. The J-K flip-flop (a) and its truth table (b). The truth table can be shortened (c) by making use of X to mean 'don't care' — either 0 or 1

edge of the clock. The use of this system ensures a controllable time delay between the output and the input, which permits signals to be fed back from one flip-flop to another without the risk of instability, since an output signal fed back to a J or K input cannot have any effect until the clock leading edge, and will not itself change again until the clock trailing edge. This avoids the difficulty of having an input changing while the output which it controls also changes.

The remaining two signal inputs of the J-K flip-flop operate independently of the clock. The SET (or PRESET) terminal sets the Q output to logic 1, regardless of other signals, and the RESET or CLEAR terminal sets the output to logic 0, also regardless of other signals. The system must be arranged so that these two cannot operate together.

Figure 5.12 shows a type of binary counter variously named serial, ripple, or asynchronous. Using four stages of J-K flip-flops, the counter Q outputs will reach 1111 (decimal 15) before resetting. By using gates to detect 1010 (decimal ten), the counter can be forced to reset on the tenth input pulse, so that a decimal count is obtained. Such a set of gated flip-flops forms the basis of a BCD (Binary-Coded Decimal) counter such as the 7490.

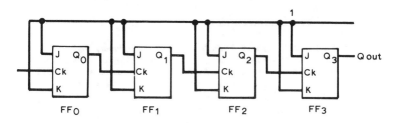

Figure 5.12. A ripple or asynchronous counter using J-K flip-flops. Note that the flip-flops (and the Q outputs) are labelled 0, 1, 2, 3 rather than 1, 2, 3, 4. This is because the first stage counts 2^0, the second stage 2^1, the third stage 2^2 and so on

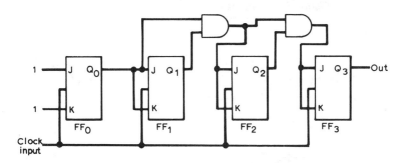

Figure 5.13. A synchronous counter. The input (clock) pulses are taken to each clock terminal, and the changeover of each flip-flop is controlled by the settings of the J, K terminals, which are gated

Figure 5.13 shows a synchronous counter, in which each flip-flop is clocked at the same rate, and counting of the clock pulses is achieved by using gated connections to the J and K inputs. The design of such synchronous counters for numbers which are not powers of two is complex and lengthy. The use of a technique called Karnaugh mapping simplifies the procedure to some extent for small count numbers, but is beyond the scope of this book. In any case, either ripple or synchronous counters can now be obtained as complete integrated circuits for most useful count values, making the design procedure unnecessary, and for complex applications, the use of a microprocessor is probably a simpler solution.

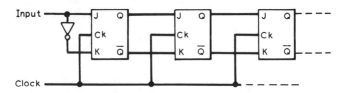

Figure 5.14. Shift register. At each clock pulse, the binary digit at the Q output of each flip-flop is 'shifted' to the next flip-flop in line. The arrangement shown here gives right-shift; left-shift can be arranged by connecting Q̄ to J and Q to K between flip-flops. For details and circuits, see Beginner's Guide to Digital Electronics

All counter circuits can be arranged so as to count in either direction up or down, and complete counters in i.c. form can be obtained which will change counting direction by altering a control voltage to 0 or 1. An example of such a counter is the 74192.

Shift registers, which can be formed from J-K flip-flops connected as shown in *Figure 5.14*, are also obtained in i.c. form. The action of a

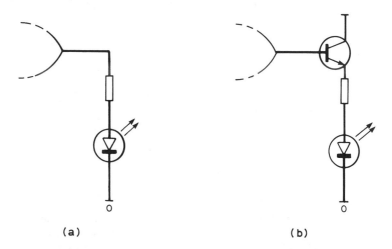

(a) (b)

Figure 5.15. LED displays. TTL i.c.s can drive the LED directly (a) using a limiting resistor. CMOS i.c.s generally have low current outputs, and a transistor 'interface' circuit (b) is needed. Multiple transistors for this purpose can be obtained in i.c. form, such as the RCA CA8083 (5 NPN) and CA3082 (7 NPN transistors with a common collector connection)

shift register is to pass a logic signal (1 or 0) from one flip-flop to the next in line at each clock pulse. The input signals can be serial, so that one bit is shifted in at each clock pulse, or parallel, switched into each flip-flop at the same time. The output can similarly be serial, taken

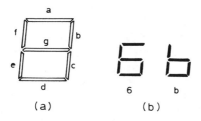

(a)   (b)

Count	Segments illuminated
O	a b c d e f
1	b c
2	a b g e d
3	a b g c d
4	f g b c
5	a f g c d
6	a f e d c g
7	a b c
8	a b c d e f g
9	a b g f c
A	a b c f e
B	f e g c d
C	a f e d
D	g e d c b
E	a f g e d
F	a f g e

(c)

Figure 5.16. The seven-segment display (a) showing the segment lettering. An eighth segment, the decimal point, is often added, and may be on the right or left side. Hexadecimal letters as well as figures can be displayed provided that 6 and b are differentiated as shown (b) by the use of the a-segment. The truth table (c) shows the segments illuminated for each number in a hex count. Displays may be common anode or common cathode

from one terminal at each clock pulse, or parallel at each flip-flop output. The shift can be left, right or switched in either direction. Some circuits which make use of shift registers are discussed in detail in *Beginner's Guide to Digital Electronics.*

Displays and decoders

A binary number can be displayed using any visible on/off device; the most convenient is usually the LED (*Figure 5.15*). For decimal number and letter displays (alphanumeric displays), dot matrix displays are used. These consist of an array of LED dots, usually in a 7 × 5 arrangement, which can be illuminated to display a number or letter. A more common arrangement is the seven-segment display (*Figure 5.16*), consisting of bars which can be illuminated. Seven-segment displays may be common-anode, with a common positive connection and each bar illuminated by connecting its terminal to logic zero. The alternative is the common-cathode type, in which the common cathode terminal is earthed and each bar anode is illuminated by taking its voltage to logic 1. A current-limiting resistor must be wired in series with each bar terminal whichever type of system is used.

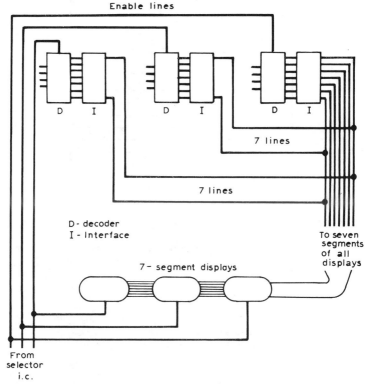

Figure 5.17. A strobed or multiplexed display. The decoder i.c.s are operated in sequence by the enable lines (from a counter), which also select the correct display. Only seven segment connections are needed for all the displays, so that fewer pins are needed. The interface i.c.s must be tri-state so that when one is active, driving the segment lines, the others are not affected

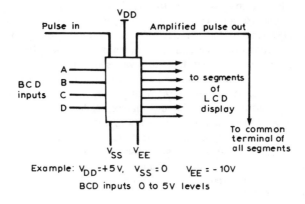

Example: $V_{DD} = +5V$, $V_{SS} = 0$ $V_{EE} = -10V$
BCD inputs 0 to 5V levels

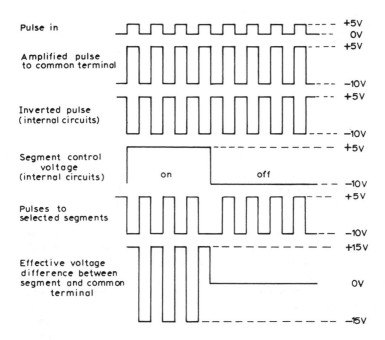

Figure 5.18. An LCD driving circuit and operating pulses. No d.c. must be applied to an LCD display, so that the pulses must be symmetrical about earth. The pulses are applied in phase to segments and common terminal (backplate) when the segments are inactive, but in antiphase to selected segments when a display is wanted

Whether the display is dot-matrix or seven-segment, a decoder i.c. is needed to convert from binary or decimal signals into the correct arrangement of bars or dots. A very common arrangement is to use binary coded decimal (BCD) in which a set of four bits represents each decimal digit, so that BCD-to-seven segment decoders are available; combined counter-decoders are also available. A few hexadecimal decoders exist for seven-segment drive, and Texas make combined decoder-display units. For dot-matrix displays, a read-only memory is usually used in place of a decoder, to take advantage of the much greater range of alphanumeric characters which can be obtained.

In some circuits, to save space or battery current, displays are strobed, meaning that the figures to be displayed are decoded and displayed one after the other, each in its correct place on the display (*Figure 5.17*).

Liquid crystal displays (LCD) are also used in battery-operated equipment. Their advantages include very low power consumption and better visibility under bright lighting, but at the expense of greater circuit complexity, higher cost, and the need for illumination in poor light. The signals from the decoder must be a.c., with no trace of d.c., at a frequency of a few hundred Hz. A typical LCD display circuit is shown in *Figure 5.18*.

I.C. types

The two main digital i.c. manufacturing methods are described as TTL (Transistor Transistor Logic) and MOS (Metal-Oxide-Silicon) — various types of MOS such as CMOS (complementary p and n channel used), PMOS (positive channel) and NMOS (negative channel) exist. TTL i.c.s use bipolar transistors in integrated form, with input and output circuits which are similar to those shown in *Figure 5.19*. Since the input is always to an emitter, the input resistance is low, and because the base of the input transistor is connected to the +5 V line, the input passes a current of about 1.6 mA when the input voltage is earth, logic 0. If an input is left unconnected, it will 'float' to logic 1, but can be affected by signals coupled by stray capacitance, so that such an input would normally be connected to +5 V through a 1 kΩ resistor. At the output, a totem-pole type of circuit is used. This can supply a current to a load which is connected between the output and earth (current sourcing), or can absorb a current from a load connected between the output and the +15 V line (current sinking). The normal TTL output stage can source or sink 16 mA, enabling it to drive up to ten TTL inputs. In the language of digital designers, the output stage has a fanout of ten.

TTL i.c.s which use these output stages must never be connected with the outputs of different units in parallel, since with one output stage at logic 1 and another at logic 0, large currents could pass,

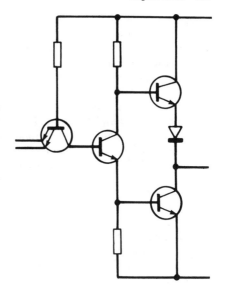

Figure 5.19. TTL circuitry. The example shown is a 2-input NAND gate. The inputs are to emitters of a transistor (in this example, a transistor with two emitters formed on to one base). The output is from a series transistor circuit so that rise and fall times are short

destroying the output stages. Modified output stages, which have open collector outputs, are available for connecting in parallel; an application which is called a wired OR, since the parallel connections create an OR gate at the output.

Inputs such as SET and RESET (or PRESET and CLEAR) normally have to be taken to logic 0 to operate, this is sometimes indicated by a small circle on the input at the symbol for the device. Clock pulses may affect the circuit either on the leading edge, trailing edge, both edges, or at logic 1; the characteristic sheet for the device must be consulted to make certain which clocking system is in use. The supply

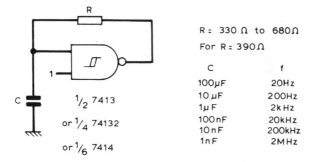

R = 330 Ω to 680Ω
For R = 390Ω

C	f
100µF	20Hz
10 µF	200Hz
1µ F	2kHz
100nF	20kHz
10nF	200kHz
1nF	2MHz

Figure 5.20. Using a Schmitt trigger i.c. as an oscillator. Any inverting Schmitt circuit can be used. The table shows very approximately the values of frequency obtained using a 390 ohm resistor and various capacitor values

voltage must be +5.0 V ± 0.25 V, and large operating currents are needed if more than a few TTL i.c.s are in use. Each group of five TTL i.c.s should have its power supply lines decoupled to avoid transmitting pulses back through the supply lines (even when a stabilised supply is used). The operating pulses should have fast rise and fall times, of the order of 50 ns or less, since many TTL circuits can oscillate if they are linearly biased, as can happen momentarily during a slow rising or falling input. A useful tip is to use Schmitt trigger gates at the input to each circuit, so ensuring a fast rising and falling pulse from any input. These circuits can also be used as clock-pulse generators (*Figure 5.20*).

Table 5.4 (p. 142) shows the pin arrangements for a number of TTL i.c.s.

Most of the TTL range are now available as low power Schottky types. These i.c.s make use of a component called the Schottky diode, which can be built in i.c. form, and which limits the voltage between the collector and the base of a transistor when connected between these points. The use of Schottky diodes makes it possible to design gates which need much smaller currents than conventional TTL i.c.s and which, because the transistors never saturate, can switch as fast, or even faster. The typical input current which has to be sunk to keep a low-power Schottky gate input at logic 0 is around 0.4 mA, only a quarter of the amount needed for a conventional stage. These Schottky i.c.s are generally distinguished by the use of the letters LS in the type number, such as 74LS00, 74LS132, etc.

MOS circuits

Small-scale digital circuits, as well as large-scale microprocessor or calculator chips, can be made using MOS techniques, making use of n-channel, p-channel, or both. One very useful series, the 4000 series, uses both p and n channel FETs, and is known as CMOS (Complementary MOS). Typical input and output circuits are shown in *Figure*

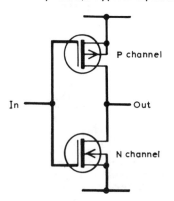

Figure 5.21. Typical CMOS input and output circuits — the example is of an inverter. In practice, protection diodes are built in at the gate inputs to prevent excessive voltages from damaging the gates, so that nothing short of a visible spark at the input is likely to cause harm

5.21. Since the input is always to the gate of a FET, the input resistance is always very high, so that virtually no d.c. will flow in the input circuit, except through protective diodes. This feature makes the fanout capability for low frequency signals very large, though at higher operating speeds the fanout is limited by the need for the output currents to charge and discharge the stray capacitances at the gate inputs. Because of the high input resistance, the input gates are easily damaged by electrostatic charges. When the i.c. is in circuit, load resistors, along with built-in protection diodes, guard against overloads, but when the i.c. is not in a circuit protection against electrostatic damage can be assured only while all the pins are shorted together. For storage, the pins are often embedded in conductive foam, or the i.c. contained in a conductive plastic case. Particular care has to be taken when CMOS i.c.s are connected into circuit. In 'normal' domestic surroundings, the following precautions are adequate.

(1) The remainder of the circuit should be complete before the MOS i.c.s are added.

(2) Plugging into holders is less hazardous than soldering.

(3) The negative line of the circuit should be earthed when the i.c.s are plugged in.

(4) If soldering is used, the soldering iron must have an earthed bit.

(5) No input pins must ever be left unconnected in a circuit.

(6) The pins of the i.c. should not be handled.

For industrial conditions where low humidity and large insulating areas can present unusual electrostatic problems, the manufacturers' guides should be consulted.

The high input resistance of CMOS i.c.s makes some circuits, particularly oscillator circuits, much easier to implement, and the very low current consumption makes the use of battery operation possible for large circuits. The maximum operating voltage is +15 V, and satisfactory operation is possible at only 3 V; a safe range of operating voltages is 4 V to 12 V. The delay timer for a typical gate is greater than that of a comparable TTL i.c., but for many applications this is not important. Clock pulses with rise or fall times exceeding 5 μs should not be used.

Table 5.5 (p. 177) shows the pin arrangement of a range of CMOS i.c.s.

Computers

The main circuit components of a digital computer of any size are registers and gates. The registers are, in the main, parallel in, parallel out types, but with provision for serial shifting. The digital computer

32		33	!	34	"	35	#	36	$
37	%	38	&	39	'	40	(	41	)
42	*	43	+	44	,	45	–	46	.
47	/	48	0	49	1	50	2	51	3
52	4	53	5	54	6	55	7	56	8
57	9	58	:	59	;	60	<	61	=
62	>	63	?	64	@	65	A	66	B
67	C	68	D	69	E	70	F	71	G
72	H	73	I	74	J	75	K	76	L
77	M	78	N	79	O	80	P	81	Q
82	R	83	S	84	T	85	U	86	V
87	W	88	X	89	Y	90	Z	91	[
92	\	93	]	94	^	95	_	96	`
97	a	98	b	99	c	100	d	101	e
102	f	103	g	104	h	105	i	106	j
107	k	108	l	109	m	110	n	111	o
112	p	113	q	114	r	115	s	116	t
117	u	118	v	119	w	120	x	121	y
122	z	123	{	124	¦	125	}	126	~
127	■								

Figure 5.22. The ASCII codes as they appear on a printer. Computers often use minor variations for some codes, particularly 123 to 127. There is no standard £ sign, for example, in the list

operates on binary numbers, and any information that is stored or used must be in binary form. Letters of the alphabet, for example, are coded as numbers, using the ASCII code system which is illustrated in *Figure 5.22.* Denary numbers are converted to binary form, using two different methods. One method is used for integers (whole numbers), and employs the methods that we have illustrated in *Table 5.1*. Numbers which are integers are stored in exact form, and the result of any arithmetic operation on an integer, with the exception of division, is also exact. The reason for excepting division is that division can result in a fraction, and integer numbers do not include fractions. Dividing the integer 5 by the integer 2, for example, gives the integer 2, because there is no way of expressing 0.5 in integer terms.

To avoid the limitations that are imposed by the use of integers, then, computers also allow for what are called 'real' or 'floating-point' numbers. This allows a large range of numbers, positive or negative, whole or fractional, to be used. The storage method is not so straightforward, however. Each real number is converted into a binary number, which can include a binary fraction. The number is then rearranged by shifting the binary point into a binary fraction and a power of two. These two are then stored and used for arithmetic. One common scheme is to store the binary fraction as a 32-bit number, and the power of two in eight bits. The snag in this scheme is that very few real numbers convert *exactly* into binary fraction form. This causes errors when real numbers are used, and the programming of the computer must be arranged so as to correct these errors by rounding numbers up or down where necessary.

The main computing actions then consist of operations on binary bits. These bits are stored in the memory of the computer, a set of miniature flip-flops or capacitors. Flip-flop memory is known as 'static' memory, and is arranged so that each group of flip-flops can be connected to by placing a binary-code address number on a set of lines. This method allows any part of the memory to be used either to store bits (writing) or to copy existing bits (reading). Memory like this is known as 'random-access memory (RAM)' to distinguish it from the older scheme which used large shift registers with the bits fed out serially. The form of RAM which uses miniature capacitors (formed as part of a MOS chip) is called 'dynamic RAM', and it uses the same principles of addressing. It is easier to construct, particularly if large memory capacity is needed, but retains data for only a matter of milliseconds. However, by arranging for the memory to be scanned at intervals and all charged capacitors recharged, the memory can be retained for as long as this scanning or 'refreshing' is applied.

The computing actions consist of addition and subtraction of binary numbers, gating under AND, OR or XOR rules, copying binary numbers from one memory address to another, and shifting or rotation actions within registers. All of these actions can be carried out by using gates and registers, and the main actions of logic gating and arithmetic are carried out by a section of the computer which is called the ALU, arithmetic and logic unit. These very simple actions are, remarkably enough, the basis of all computing operations.

Programmability

The feature that makes the computer so useful is programmability. When we construct a gate and register circuit, the circuit paths are normally fixed, and we can change them only by cutting tracks on the PCB and resoldering. Imagine, however, a circuit which contained gates and registers, but in which all the circuit paths could be changed by using gates in each connection (*Figure 5.23*). In the simplified example, the signal at S can be routed to register 1 or to register 2 according to the control voltage C. If C is at logic 1, then the signal at S reaches register 1.

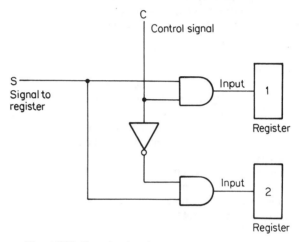

Figure 5.23. How signal paths can be controlled by gates

If the control voltage at C is at logic 0, then the signal at S reaches register 2. If we provide enough of these gates, all of the signal paths in a circuit can be changed over simply by connecting a suitable logic voltage to a set of gates.

Now we can take this idea a step further. Suppose that the gates which control the connections are themselves operated by the bits stored in a register. By changing the bits that are stored in this register, we can change the connections in a circuit, and so alter the action of the circuit. This is how a circuit can be made programmable, and the program consists of the pattern of bits that is stored in the register that controls the internal circuit gates. In a computer, the register that is used in this way is called the 'instruction register'. All of the actions of the ALU, and of the other computer circuits, can be controlled by the binary number bits that are placed in this instruction register. In practice, each little section is controlled by its own instruction register, and all of these minor instruction registers are controlled by one main instruction register. The sets of numbers which are used in the minor instruction registers are called the 'microprogram' for the system. This whole set-up is known as the CPU, central processing unit, of the computer.

The action of the computer is then controlled by feeding numbers into the instruction register. One set of numbers sent to the instruction register, for example, may cause connections to be made so that a number is copied from memory, added to another number, and then stored back in memory. Another set may cause a number to be read from memory, shifted left, then stored back in memory. All of these operations are programmed by making use of the instruction register. The design of the registers is such that all actions are clocked, and the clock rate can be as fast as the type of semiconductors permits. If, for example, the clock rate is 1 MHz, then one million internal actions per second can be performed. This is *not* the same as one million additions per second, because one addition requires several internal actions of copying and gating to be carried out. A reasonable average is to assume that each computer operation will take from 3 to 12 clock pulses to complete. This is still a fairly fast rate, particularly when we consider that a 1 MHz clock rate is hopelessly slow by modern standards. Even microcomputers commonly use 4 MHz clock rates, and larger computers use rates of 100 MHz or more.

The 'smart' actions of computers are achieved by programming, feeding binary command numbers into the main instruction register as fast as they are needed. This cannot be done manually, because of the speed of the clock, and the method that is used is to store the correct sequence of command numbers in memory. The computer then reads the numbers as and when they are needed. This is possible because the memory is 'addressable', meaning that any part of the memory can be reached by putting a binary number in bit form on to a set of lines. These lines, called the address bus, are connected to all of the memory units, and another set of lines, the data bus, carry the signals which form the bits of numbers that the computer is working on, the data. The address bus obtains its signals from a register, the 'program counter', and the action of the CPU is that the program counter number is increased by one step each time a program command number has to be used. If these command numbers are stored in *exactly* the correct sequence in the memory, then, the whole action is automatic. If the CPU program counter register is loaded with the memory address of the first number of a program, this number will be copied into the instruction register, and the automatic action will start. If the instruction requires another number to be read, the program counter increments (its content has 1 added to it), and this next memory address is read. At the end of the first instruction, the program counter increments again to read the next instruction code. This will continue until one of the codes causes the CPU to stop. Even this term is misleading, because the presence of the clock pulses means that the CPU cannot stop. The nearest it can come to stopping is to enter a 'loop'. This means executing the same instructions over and over again until something is done to start another program running. The instructions can be stored in the ordinary RAM memory, in which case they will be lost when power is switched off. An alternative is to store instructions in a different type of memory, ROM. ROM means 'read-only memory', and such a memory consists of a set of permanent connections, some to earth (logic 0) and some to a power supply pin (logic 1). Once again, these connections are reached by making use of address signals, so that for each address number, a different set of instruction bits is available. The advantage of using ROM is that the instructions are not lost when power is switched off, because the connections are permanent. The disadvantage, however, is that the ROM cannot be altered if another program is needed.

This brief description of CPU action nevertheless points out one very important feature. The CPU requires for its operation a set of binary numbers which act as its programming instructions. These numbers have to be stored in RAM (or ROM) memory, and they have to be stored in *exactly* the order that the CPU will need them. All the actions that are achieved by the computer are due to programs of this type stored in memory. Remember that the fundamental actions of the CPU may be no more than addition, subtraction, logic gating, and shifting. Anything else that the CPU can accomplish has to be done by a program in the memory. The set of number codes that dictates the action of the CPU is called 'machine-code', and all programming of computers has ultimately to be done in machine-code.

Each CPU design requires different machine code instructions, and writing directly in machine code is a very tedious business, mainly because of the amount of detail that is needed. There may, for example, be several hundred steps of machine code needed for an action like extracting a square root of a number. Because this type of programming is so tedious, *high-level* languages have been devised. A high level language consists of a set of instruction words and a syntax. The instruction words are like ordinary English words, words such as PRINT, DO, REPEAT, UNTIL and so on. The 'syntax' means the way in which the words are used. A computer which can use a high-level language contains a program, written in machine code, which can interpret the meaning of the instruction words and convert a program written in this way into machine code. This can be done in two ways. One way is called 'interpretation', and this is the method that is used by most of the very small microcomputers. In an interpreter, each instruction word, and the things it operates on (numbers, letters, words) will be converted into a set of machine-code numbers, and the machine code is run. This is done immediately after an instruction is interpreted, and the next instruction is then converted into machine code and run. The conversion process takes time — it consists of reading the instruction word, and looking for a matching word in the part of the memory that holds the interpreter program. When a match is found, this leads to an address for the machine-code that carries out the action. All of this code is already stored as part of the interpreter, but finding it takes time. This means that an interpreted program runs comparatively slowly, because the looking-up process for each instruction takes time, and the instruction cannot be carried out until the looking-up is completed.

The alternative is called 'compiling'. The instructions are read, looked up, and translated into machine code as before. This time, though, each section of machine code is recorded, to form a complete machine-code program. Once this compiling action has been done, the recorded machine code program can be put into the memory and run at any time. This program will run fast, because it does not have to wait for any looking up processes to be completed. All of the looking up actions have been done once during compiling, and do not have to be done again.

Pinouts

The following pages illustrate labelled pinout diagrams for the two most common 8-bit microprocessors, the 6502 and the Z80, and a pinout diagram for the 16-bit 68010. Since all microprocessor systems are clocked, the purity of the clock signal is very important, and all of the timings assume that the clock signal will be within specification limits. Servicing work on microprocessor equipment is made much easier if specialised logic analysers are available. A good storage oscilloscope is also very useful in order to show the time relationships of signals which may not occur frequently enough to trigger a conventional display.

D4	1●	64	D5
D3	2	63	D6
D2	3	62	D7
D1	4	61	D8
D0	5	60	D9
$\overline{AS}$	6	59	D10
$\overline{UDS}$	7	58	D11
$\overline{LDS}$	8	57	D12
$R/\overline{W}$	9	56	D13
$\overline{DTACK}$	10	55	D14
$\overline{BG}$	11	54	D15
$\overline{BGACK}$	12	53	GND
$\overline{BR}$	13	52	A23
V_{CC}	14	51	A22
CLK	15	50	A21
GND	16	49	V_{CC}
$\overline{HALT}$	17	48	A20
$\overline{RESET}$	18	47	A19
$\overline{VMA}$	19	46	A18
E	20	45	A17
$\overline{VPA}$	21	44	A16
$\overline{BERR}$	22	43	A15
$\overline{IPL2}$	23	42	A14
$\overline{IPL1}$	24	41	A13
$\overline{IPL0}$	25	40	A12
FC2	26	39	A11
FC1	27	38	A10
FC0	28	37	A9
A1	29	36	A8
A2	30	35	A7
A3	31	34	A6
A4	32	33	A5

68010

	Z 80	
A11 — 1		40 — A10
A12 — 2		39 — A9
A13 — 3		38 — A8
A14 — 4		37 — A7
A15 — 5		36 — A6
CLK — 6		35 — A5
D4 — 7		34 — A4
D3 — 8		33 — A3
D5 — 9		32 — A2
D6 — 10		31 — A1
+5V — 11		30 — A0
D2 — 12		29 — GND
D7 — 13		28 — RFSH
D0 — 14		27 — MI
D1 — 15		26 — RESET
INT — 16		25 — BUSREQ
NMI — 17		24 — WAIT
HALT — 18		23 — BUSAK
MREQ — 19		22 — WR
IORQ — 20		21 — RD

	6502	
Vss — 1		40 — $\overline{\text{RESET}}$
RIY — 2		39 — $\phi2$(OUT)
$\phi1$(OUT) — 3		38 — SO
$\overline{\text{IRQ}}$ — 4		37 — $\phi0$(IN)
NC — 5		36 — NC
$\overline{\text{NMI}}$ — 6		35 — NC
SYNC — 7		34 — R/$\overline{\text{W}}$
+5V — 8		33 — D0
A0 — 9		32 — D1
A1 — 10		31 — D2
A2 — 11		30 — D3
A3 — 12		29 — D4
A4 — 13		28 — D5
A5 — 14		27 — D6
A6 — 15		26 — D7
A7 — 16		25 — A15
A8 — 17		24 — A14
A9 — 18		23 — A13
A10 — 19		22 — A12
A11 — 20		21 — Vss

Table 5.4. TTL PINOUTS

7400
QUADRUPLE 2-INPUT
POSITIVE-NAND GATES

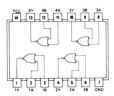

7401
QUADRUPLE 2-INPUT
POSITIVE-NAND GATES
WITH OPEN-COLLECTOR OUTPUTS

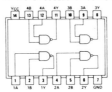

74H01

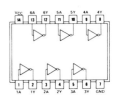

7402
QUADRUPLE 2-INPUT
POSITIVE-NOR GATES

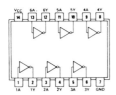

7403
QUADRUPLE 2-INPUT
POSITIVE-NAND GATES
WITH OPEN-COLLECTOR OUTPUTS

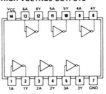

7404
HEX INVERTERS

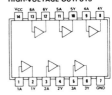

7405
HEX INVERTERS
WITH OPEN-COLLECTOR OUTPUTS

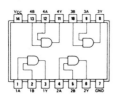

7406
HEX INVERTER BUFFERS/DRIVERS
WITH OPEN-COLLECTOR
HIGH-VOLTAGE OUTPUTS

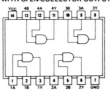

7407
HEX BUFFERS/DRIVERS
WITH OPEN-COLLECTOR
HIGH-VOLTAGE OUTPUTS

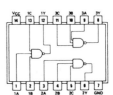

7408
QUADRUPLE 2-INPUT
POSITIVE-AND GATES

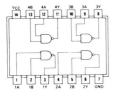

7409
QUADRUPLE 2-INPUT
POSITIVE-AND GATES
WITH OPEN-COLLECTOR OUTPUTS

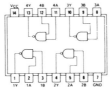

7410
TRIPLE 3-INPUT
POSITIVE-NAND GATES

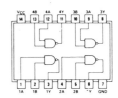

74H11
TRIPLE 3-INPUT
POSITIVE-AND GATES

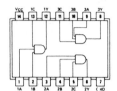

7412
TRIPLE 3-INPUT
POSITIVE-NAND GATES
WITH OPEN-COLLECTOR OUTPUTS

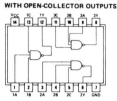

7413
DUAL 4-INPUT
POSITIVE-NAND
SCHMITT TRIGGERS

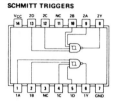

7414
HEX SCHMITT-TRIGGER
INVERTERS

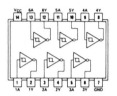

74H15
TRIPLE 3-INPUT
POSITIVE-AND GATES
WITH OPEN-COLLECTOR OUTPUTS

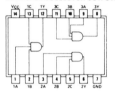

7416
HEX INVERTER BUFFERS/DRIVER
WITH OPEN-COLLECTOR
HIGH-VOLTAGE OUTPUTS

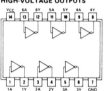

7417
HEX BUFFERS/DRIVERS
WITH OPEN-COLLECTOR
HIGH-VOLTAGE OUTPUTS

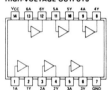

7420
DUAL 4-INPUT
POSITIVE-NAND GATES

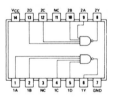

74H21
DUAL 4-INPUT
POSITIVE-AND GATES

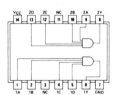

7422
DUAL 4-INPUT
POSITIVE-NAND GATES
WITH OPEN-COLLECTOR OUTPUTS

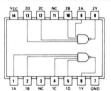

7423
EXPANDABLE DUAL 4-INPUT
POSITIVE-NOR GATES
WITH STROBE

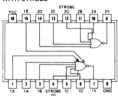

7425
DUAL 4-INPUT
POSITIVE-NOR GATES
WITH STROBE

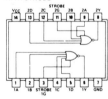

7426
QUADRUPLE 2-INPUT
HIGH-VOLTAGE INTERFACE
POSITIVE-NAND GATES

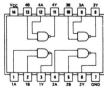

7427
TRIPLE 3-INPUT
POSITIVE-NOR GATES

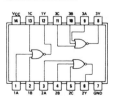

7428
QUADRUPLE 2-INPUT
POSITIVE-NOR BUFFERS

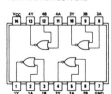

7430
8-INPUT
POSITIVE-NAND GATES

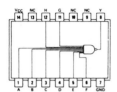

7432
QUADRUPLE 2-INPUT
POSITIVE-OR GATES

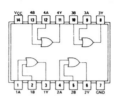

7433
QUADRUPLE 2-INPUT
POSITIVE-NOR BUFFERS
WITH OPEN-COLLECTOR OUTPUTS

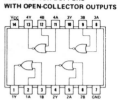

7437
QUADRUPLE 2-INPUT
POSITIVE-NAND BUFFERS

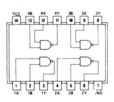

7438
QUADRUPLE 2-INPUT
POSITIVE-NAND BUFFERS
WITH OPEN-COLLECTOR OUTPUTS

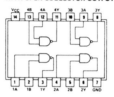

7440
DUAL 4-INPUT
POSITIVE-NAND BUFFERS

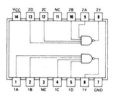

7442A
4 LINE-TO-10-LINE DECODERS

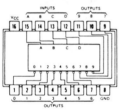

7445
BCD-TO-DECIMAL DECODER/DRIVER

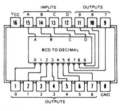

7446
BCD-TO-SEVEN-SEGMENT
DECODERS/DRIVERS

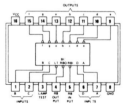

7448
BCD-TO-SEVEN-SEGMENT
DECODERS/DRIVERS

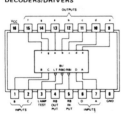

74LS49
BCD-TO-SEVEN-SEGMENT
DECODERS/DRIVERS

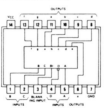

7450
DUAL 2-WIDE 2-INPUT
AND-OR-INVERT GATES
(ONE GATE EXPANDABLE)

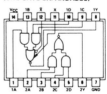

7451
AND-OR-INVERT GATES
MAKE NO EXTERNAL CONNECTION

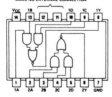

74L51

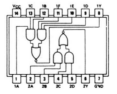

74H52
EXPANDABLE 4-WIDE
AND-OR GATES

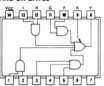

7453
EXPANDABLE 4-WIDE
AND-OR-INVERT GATES

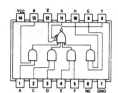

74H53

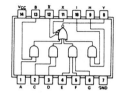

7454
4-WIDE
AND-OR-INVERT GATES

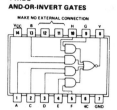

74H54

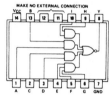

74L54

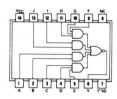

74H55
2-WIDE 4-INPUT
AND-OR-INVERT GATES

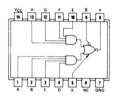

74L55

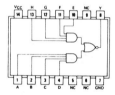

7460
DUAL 4-INPUT EXPANDERS

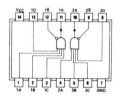

74H61
TRIPLE 3-INPUT
EXPANDERS

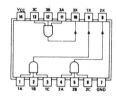

74H62
4-WIDE AND-OR EXPANDERS

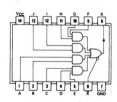

74LS63
HEX CURRENT-SENSING
INTERFACE GATES

74S64
4-2-3-2 INPUT AND-OR-INVERT
GATES

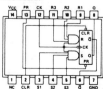

7470
AND-GATED J-K POSITIVE-EDGE-
TRIGGERED FLIP-FLOPS
WITH PRESET AND CLEAR

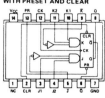

74H71
AND-OR-GATED J-K MASTER-SLAVE
FLIP-FLOPS WITH PRESET

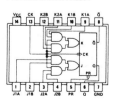

74L71
AND-GATED R-S MASTER-SLAVE
FLIP-FLOPS WITH PRESET
AND CLEAR

/472
AND-GATED J-K MASTER-SLAVE FLIP-FLOPS WITH PRESET AND CLEAR

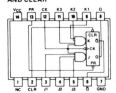

7473
DUAL J-K FLIP-FLOPS WITH CLEAR

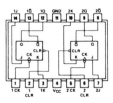

7474
DUAL D-TYPE POSITIVE-EDGE-TRIGGERED FLIP-FLOPS WITH PRESET AND CLEAR

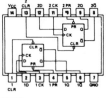

7475
4-BIT BISTABLE LATCHES

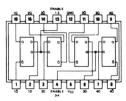

7476
DUAL J-K FLIP-FLOPS WITH PRESET AND CLEAR

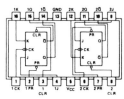

74H78
DUAL J-K FLIP-FLOPS WITH PRESET, COMMON CLEAR, AND COMMON CLOCK

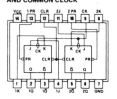

74L78

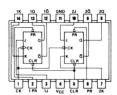

7480
GATED FULL ADDERS

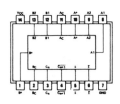

7481
16-BIT RANDOM-ACCESS MEMORIES

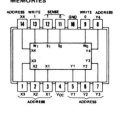

7482
2-BIT BINARY FULL ADDERS

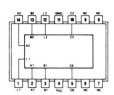

74 83
4-BIT BINARY FULL ADDERS WITH FAST CARRY

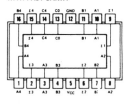

7484
16-BIT RANDOM-ACCESS MEMORIES

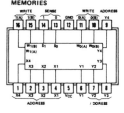

7485
4-BIT MAGNITUDE COMPARATORS

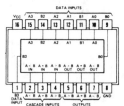

7486
QUADRUPLE 2-INPUT EXCLUSIVE-OR GATES

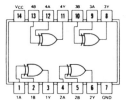

74L86

74H87
4-BIT TRUE/COMPLEMENT,
ZERO/ONE ELEMENTS

7488A
256-BIT READ-ONLY MEMORIES

7489
64-BIT READ/WRITE MEMORIES

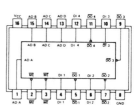

7490A
DECADE COUNTERS

7491A
8-BIT SHIFT REGISTERS

7492
DIVIDE-BY-TWELVE COUNTERS

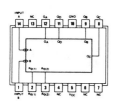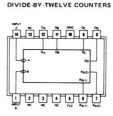

7493A
4-BIT BINARY COUNTERS

7494
4-BIT SHIFT REGISTERS

7495A
4-BIT SHIFT REGISTERS

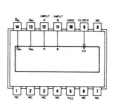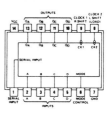

7496
5-BIT SHIFT REGISTERS

7497
SYNCHRONOUS 6-BIT BINARY
RATE MULTIPLIERS

74L98
4-BIT DATA SELECTOR/STORAGE
REGISTERS

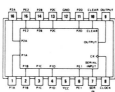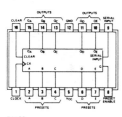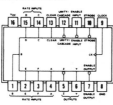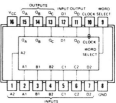

74L99
4-BIT BIDIRECTIONAL UNIVERSAL
SHIFT REGISTERS

74100
8-BIT BISTABLE LATCHES

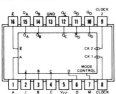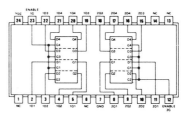

74H101
AND-OR-GATED J-K NEGATIVE-EDGE-TRIGGERED FLIP-FLOPS WITH PRESET

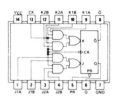

74H102
AND-GATED J-K NEGATIVE-EDGE-TRIGGERED FLIP-FLOPS WITH PRESET AND CLEAR

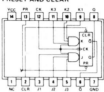

74H103
DUAL J-K NEGATIVE-EDGE-TRIGGERED FLOPS WITH CLEAR

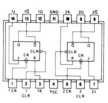

74H106
DUAL J-K NEGATIVE-EDGE-TRIGGERED FLIP-FLOPS WITH PRESET AND CLEAR

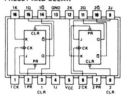

74107
DUAL J-K FLIP-FLOPS WITH CLEAR

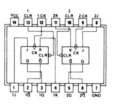

74H108
DUAL J-K NEGATIVE-EDGE-TRIGGERED FLIP-FLOPS WITH PRESET, COMMON CLEAR, AND COMMON CLOCK

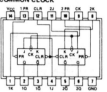

74109
DUAL J-K̄ POSITIVE-EDGE-TRIGGERED FLIP-FLOPS WITH PRESET AND CLEAR

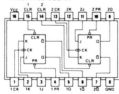

74110
AND-GATED J-K MASTER-SLAVE FLIP-FLOPS WITH DATA LOCKOUT

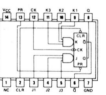

74111
DUAL J-K MASTER-SLAVE FLIP-FLOPS WITH DATA LOCKOUT

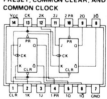

74LS112A
DUAL J-K NEGATIVE-EDGE-TRIGGERED FLIP-FLOPS WITH PRESET AND CLEAR

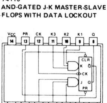

74LS113A
DUAL J-K NEGATIVE-EDGE-TRIGGERED FLIP-FLOPS WITH PRESET

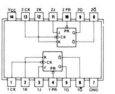

74LS114A
DUAL J-K NEGATIVE-EDGE-TRIGGERED FLIP-FLOPS WITH PRESET, COMMON CLEAR, AND COMMON CLOCK

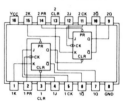

74116
DUAL 4-BIT LATCHES

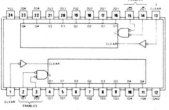

74120
DUAL PULSE SYNCHRONIZERS/DRIVERS

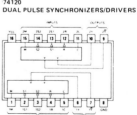

74121
MONOSTABLE MULTIVIBRATORS

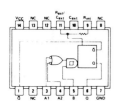

74122
RETRIGGERABLE MONOSTABLE
MULTIVIBRATORS WITH CLEAR

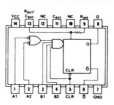

74123
DUAL RETRIGGERABLE MONOSTABLE
MULTIVIBRATORS WITH CLEAR

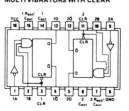

74LS124
DUAL VOLTAGE-CONTROLLED
OSCILLATORS

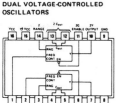

74125
QUADRUPLE BUS BUFFER GATES
WITH THREE-STATE OUTPUTS

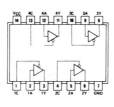

74126
QUADRUPLE BUS BUFFER GATES
WITH THREE-STATE OUTPUTS

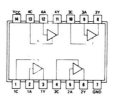

74128
SN74128 . . . 50-OHM LINE DRIVER

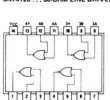

74132
QUADRUPLE 2-INPUT POSITIVE-
NAND SCHMITT TRIGGERS

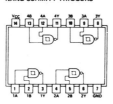

74S133
13-INPUT POSITIVE-NAND GATES

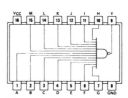

74S134
12-INPUT POSITIVE-NAND GATES
WITH THREE-STATE OUTPUTS

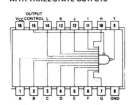

74S135
QUAD EXCLUSIVE-OR/NOR GATES

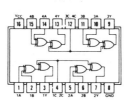

74136
QUAD EXCLUSIVE-OR GATES

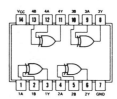

74LS138
3-TO-8 LINE DECODERS/
MULTIPLEXERS

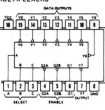

74LS139
DUAL 2-TO-4 LINE DECODERS/
MULTIPLEXERS

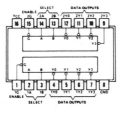

74S140
DUAL 4-INPUT POSITIVE-NAND
50-OHM LINE DRIVERS

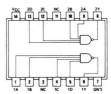

74141
BCD-TO-DECIMAL DECODER/ DRIVER

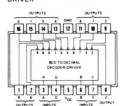

74142
COUNTER/LATCH/DECODER/ DRIVER

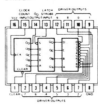

74143   74144
COUNTERS/LATCHES/DECODERS/ DRIVERS

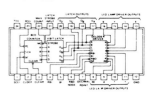

74145
BCD-TO-DECIMAL DECODERS/DRIVERS FOR LAMPS, RELAYS, MOS

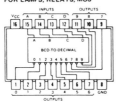

74147
10-LINE DECIMAL TO 4-LINE BCD PRIORITY ENCODERS

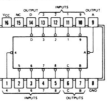

74148
8-LINE-TO-3-LINE OCTAL PRIORITY ENCODERS

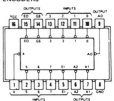

74150
1-OF-16 DATA SELECTORS/ MULTIPLEXERS

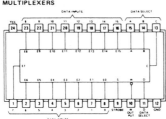

74151A
1-OF-8 DATA SELECTORS/MULTIPLEXERS

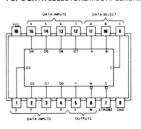

74153
DUAL 4-LINE TO 1-LINE DATA SELECTORS/MULTIPLEXERS

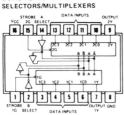

74154
4-LINE TO 16-LINE DECODERS/ DEMULTIPLEXERS

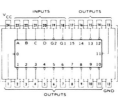

74155
DECODERS/DEMULTIPLEXERS

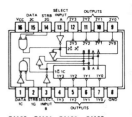

74157
QUAD 2- TO 1-LINE DATA SELECTORS/MULTIPLEXERS

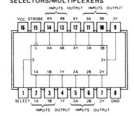

74159
4- TO 16-LINE DECODERS/ DEMULTIPLEXERS

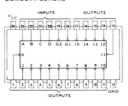

74160   74161   74162   74163
SYNCHRONOUS 4-BIT COUNTERS

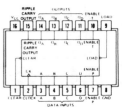

74164
8-BIT PARALLEL OUTPUT SERIAL SHIFT REGISTERS

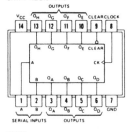

74166
8-BIT SHIFT REGISTERS

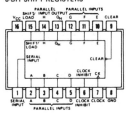

74S168   74S169
4-BIT UP/DOWN SYNCHRONOUS COUNTERS

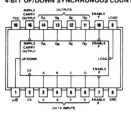

74172
16-BIT REGISTER FILE

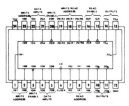

74174
HEX D-TYPE FLIP-FLOPS

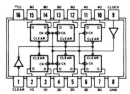

74165
PARALLEL-LOAD 8-BIT SHIFT REGISTERS WITH COMPLEMENTARY OUTPUTS

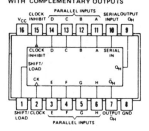

74167
SYNCHRONOUS DECADE RATE MULTIPLIERS

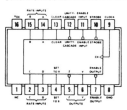

74170
4-BY-4 REGISTER FILES

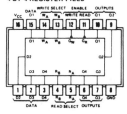

74173
4-BIT D-TYPE REGISTERS

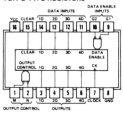

74175
QUAD D-TYPE FLIP-FLOPS

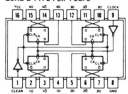

74176　74177
PRESETABLE COUNTERS/LATCHES

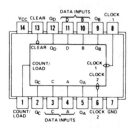

74178
4-BIT UNIVERSAL SHIFT REGISTERS

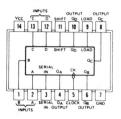

74179
4-BIT UNIVERSAL SHIFT REGISTERS

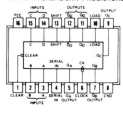

74180
9-BIT ODD/EVEN PARITY GENERATORS/CHECKERS

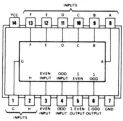

74181
ARITHMETIC LOGIC UNITS/FUNCTION GENERATORS

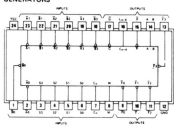

74182
LOOK-AHEAD CARRY GENERATORS

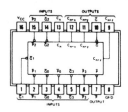

74LS183
DUAL CARRY-SAVE FULL ADDERS

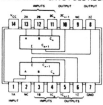

74184　74185A
CODE CONVERTERS

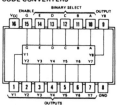

7418G
512-BIT PROGRAMABLE READ-ONLY MEMORIES

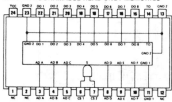

74187
1024-BIT READ-ONLY MEMORIES

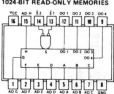

74188A
256-BIT PROGRAMMABLE READ-ONLY MEMORIES

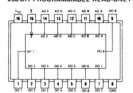

74S189
64-BIT RANDOM-ACCESS MEMORIES

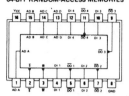

74190 74191
SYNCHRONOUS UP/DOWN COUNTERS

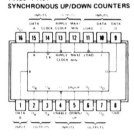

74192 74193
SYNCHRONOUS UP/DOWN DUAL CLOCK COUNTERS

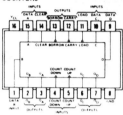

74194
4-BIT BIDIRECTIONAL UNIVERSAL SHIFT REGISTERS

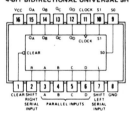

74195
4-BIT PARALLEL-ACCESS SHIFT REGISTERS

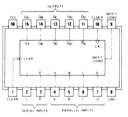

74196 74197
PRESETABLE COUNTERS/LATCHES

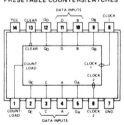

74198
8-BIT BIDIRECTIONAL UNIVERSAL SHIFT REGISTERS

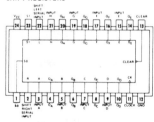

74199
8-BIT BIDIRECTIONAL UNIVERSAL SHIFT REGISTERS

74LS200A
256-BIT RANDOM-ACCESS MEMORIES

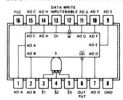

74S201
256-BIT RANDOM-ACCESS MEMORIES

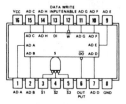

74LS207
RANDOM-ACCESS MEMORIES

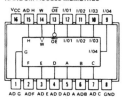

74LS214   74LS215
RANDOM-ACCESS MEMORIES

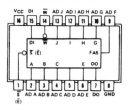

74S225
ASYNCHRONOUS FIRST IN, FIRST OUT MEMORIES

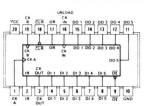

74LS240
OCTAL BUFFERS/LINE DRIVERS/LINE RECEIVERS

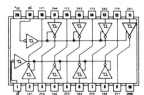

74LS202
256-BIT READ/WRITE MEMORIES WITH POWER DOWN

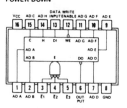

74LS208
RANDOM-ACCESS MEMORIES

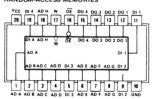

74221
DUAL MONOSTABLE MULTIVIBRATORS

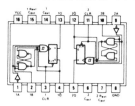

74S226
4-BIT PARALLEL LATCHED BUS TRANSCEIVERS

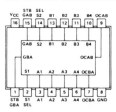

74LS241
OCTAL BUFFERS/LINE DRIVERS/LINE RECEIVERS

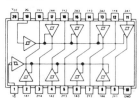

74LS242
QUADRUPLE BUS TRANSCEIVERS

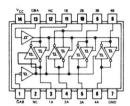

74LS243
QUADRUPLE BUS TRANCEIVERS

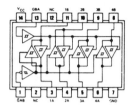

74LS244
OCTAL BUFFERS/LINE DRIVERS/LINE RECEIVERS

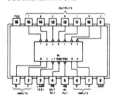

74LS245
OCTAL BUS TRANCEIVERS

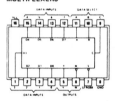

74246   74247
BCD-TO-SEVEN-SEGMENT
DECODERS/DRIVERS

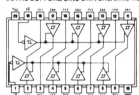

74248   74249
BCD-TO-SEVEN-SEGMENT
DECODERS/DRIVERS

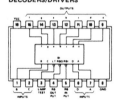

74251
DATA SELECTORS/
MULTIPLEXERS

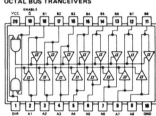

74LS253
DUAL DATA SELECTORS/MULTIPLEXERS

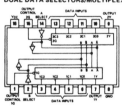

74LS257A
QUAD DATA SELECTORS/MULTIPLEXERS

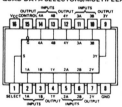

74LS258A
QUAD DATA SELECTORS/MULTIPLEXERS

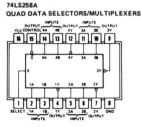

74259
EIGHT-BIT ADDRESSABLE LATCHES

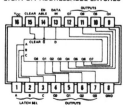

74S260
DUAL 5-INPUT POSITIVE NOR GATES

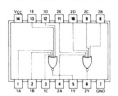

74265
QUAD COMPLEMENTARY-OUTPUT ELEMENTS

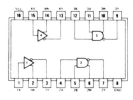

74S270
2048-BIT READ-ONLY MEMORIES

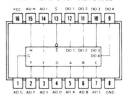

74273
OCTAL D-TYPE FLIP-FLOPS

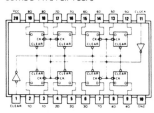

74LS275
7-BIT SLICE WALLACE TREES

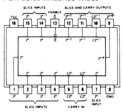

74LS261
2-BIT BY 4-BIT PARALLEL BINARY MULTIPLIERS

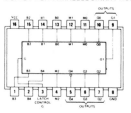

74LS266
**QUAD 2-INPUT EXCLUSIVE-NOR GATES WITH
OPEN-COLLECTOR OUTPUTS**

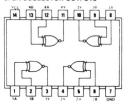

74S271
2048-BIT READ-ONLY MEMORIES

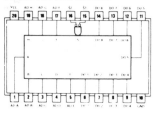

74S274
4-BIT BY 4-BIT BINARY MULTIPLIERS

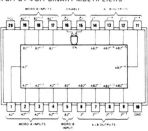

74276
QUAD J-K̄ FLIP-FLOPS

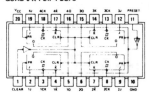

4-BIT CASCADEABLE PRIORITY REGISTERS

74279
QUAD S-R LATCHES

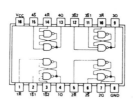

74LS280
9-BIT ODD/EVEN PARITY GENERATORS/CHECKERS

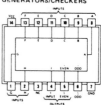

74S281
4-BIT PARALLEL BINARY ACCUMULATORS

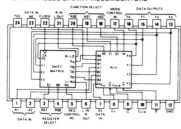

74283
4-BIT BINARY FULL ADDERS

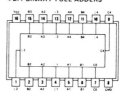

74284
4-BIT-BY-4-BIT PARALLEL BINARY MULTIPLIERS USED WITH '285

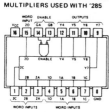

74285
4-BIT-BY-4-BIT PARALLEL BINARY MULTIPLIERS USED WITH '284

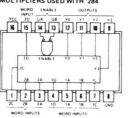

74S287
1024-BIT PROGRAMMABLE READ-ONLY MEMORIES

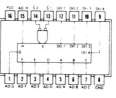

74S288
256-BIT PROGRAMMABLE READ-ONLY MEMORIES

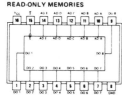

74S289
64-BIT RANDOM-ACCESS MEMORIES

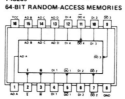

74290
DECADE COUNTERS

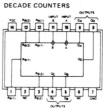

74293
4-BIT BINARY COUNTERS

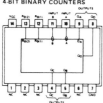

74LS295B
4-BIT BIDIRECTIONAL UNIVERSAL SHIFT REGISTERS

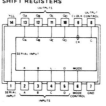

74298
QUAD 2-INPUT MULTIPLEXERS WITH STORAGE

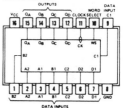

74LS299
8-BIT BIDIRECTIONAL UNIVERSAL SHIFT/STORAGE REGISTERS

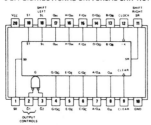

74LS300A
256-BIT READ/WRITE MEMORIES

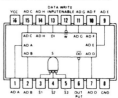

74S301
256-BIT RANDOM ACCESS MEMORIES

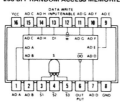

74LS302
256-BIT READ/WRITE MEMORIES

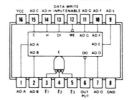

74LS314   74LS315
1024-BIT RANDOM-ACCESS MEMORIES

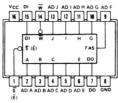

74LS323
8-BIT BIDIRECTIONAL UNIVERSAL SHIFT/STORAGE REGISTERS

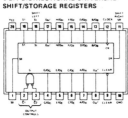

74LS324
VOLTAGE-CONTROLLED OSCILLATORS

74LS325
DUAL VOLTAGE-CONTROLLED OSCILLATORS

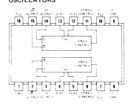

74LS326
DUAL VOLTAGE-CONTROLLED OSCILLATORS

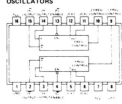

74LS327
DUAL VOLTAGE-CONTROLLED OSCILLATORS

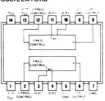

74LS348
8-LINE-TO-3-LINE PRIORITY ENCODERS

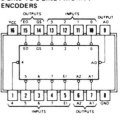

74351
DUAL 8-LINE-TO-1-LINE DATA SELECTOR/MULTIPLEXER

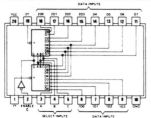

74LS352
DUAL 4-LINE-TO-LINE DATA SELECTORS/MULTIPLEXERS

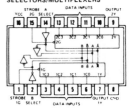

74LS353
DUAL 4-LINE-TO-1-LINE DATA SELECTORS/MULTIPLEXERS

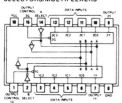

74LS362
FOUR-PHASE CLOCK GENERATOR/DRIVER FOR TMS 9900 MICROPROCESSOR

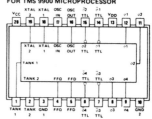

74LS363
OCTAL D-TYPE LATCHES

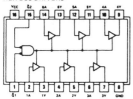

I74LS364
OCTAL D-TYPE FLIP-FLOPS

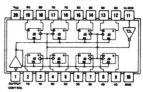

I74365A
HEX BUS DRIVERS

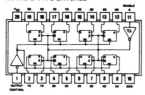

74366A
HEX BUS DRIVERS

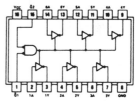

74367A
HEX BUS DRIVERS

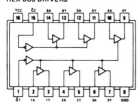

74368A
HEX BUS DRIVERS

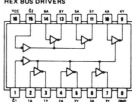

74S370
2048-BIT READ-ONLY MEMORIES

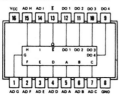

74S371
2048-BIT READ-ONLY MEMORIES

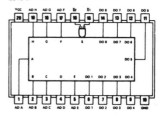

74LS373
OCTAL D-TYPE LATCHES

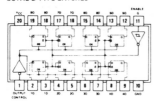

74LS374
OCTAL D-TYPE FLIP-FLOPS

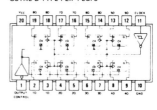

74LS375
4-BIT BISTABLE LATCHES

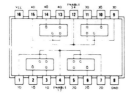

74376
QUAD J-K̄ FLIP-FLOPS

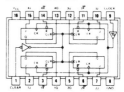

74LS377
OCTAL D-TYPE FLIP-FLOPS

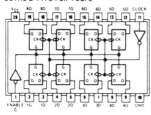

74LS378
HEX D-TYPE FLIP-FLOPS

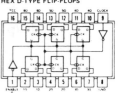

74LS379
QUAD D-TYPE FLIP-FLOPS

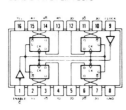

74S381
ARITHMETIC LOGIC UNITS/FUNCTION GENERATORS

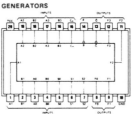

74LS386
QUAD 2-INPUT EXCLUSIVE-OR GATES

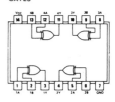

74S387
1024-BIT PROGRAMMABLE READ-ONLY MEMORIES

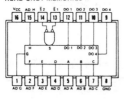

74390
DUAL DECADE COUNTERS

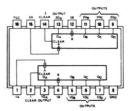

74393
DUAL 4-BIT BINARY COUNTERS

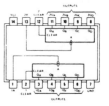

74LS395A
4-BIT UNIVERSAL SHIFT REGISTERS

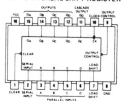

74LS398
QUAD 2-INPUT MULTIPLEXERS WITH STORAGE

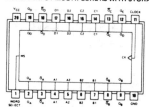

74LS399
QUAD 2-INPUT MULTIPLEXERS WITH STORAGE

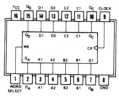

74S412
MULTI-MODE BUFFERED 8-BIT LATCHES

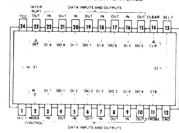

74LS424
TWO-PHASE CLOCK GENERATOR/
DRIVER FOR 8080A

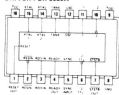

74425
QUAD GATES

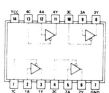

74426
QUAD GATES

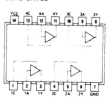

74S428   74S438
SYSTEM CONTROLLER FOR 8080A

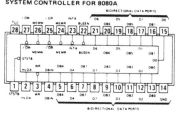

74S470   74S471
PROGRAMMABLE READ-ONLY MEMORIES

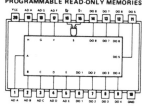

74S472   74S473
PROGRAMMABLE READ-ONLY MEMORIES

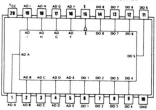

74S474   74S475
PROGRAMMABLE READ-ONLY MEMORIES

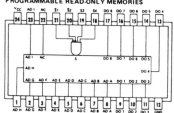

74S481
4-BIT SLICE PROCESSOR ELEMENTS

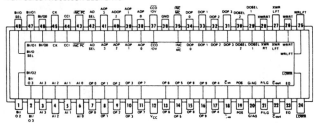

74S482
4-BIT-SLICE EXPANDABLE CONTROL ELEMENTS

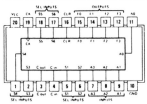

74490
DUAL DECADE COUNTERS

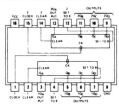

74LS670
4-BY-4 REGISTER FILES

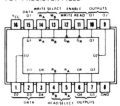

Table 5.5. CMOS PINOUTS

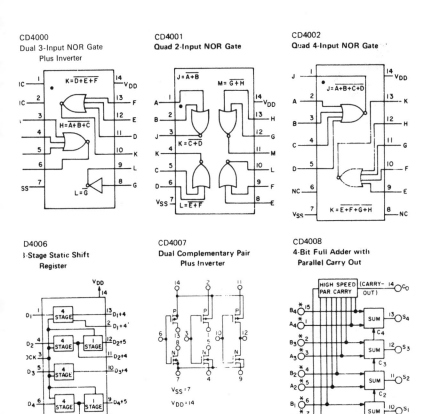

CD4000
Dual 3-Input NOR Gate
Plus Inverter

CD4001
Quad 2-Input NOR Gate

CD4002
Quad 4-Input NOR Gate

D4006
3-Stage Static Shift
Register

CD4007
Dual Complementary Pair
Plus Inverter

CD4008
4-Bit Full Adder with
Parallel Carry Out

CD4009
Hex Buffer/Converter Inverting Type

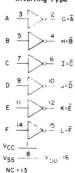

$V_{CC} = 1$
$V_{SS} = 8$ $V_{DD} = 16$
NC = 13

CD4010
Hex Buffer/Converter Non-Inverting Type

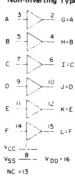

$V_{CC} = 1$
$V_{SS} = 8$ $V_{DD} = 16$
NC = 13

CD4011
Quad 2-Input NAND Gate

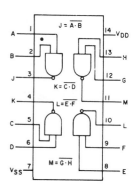

CD4012
Dual 4-Input NAND Gate

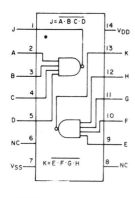

CD4013
Dual "D" Flip-Flop with Set/Reset Capability

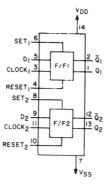

CD4014
8-Stage Synchronous Shift Register with Parallel or Serial Input/Serial Output

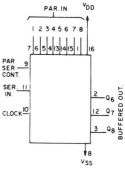

CD4015
Dual 4-Stage Static Shift Register with Serial Input/Parallel Output

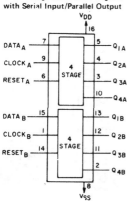

CD4016
Quad Bilateral Switch

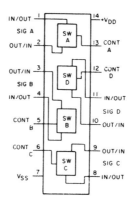

CD4017
Decade Counter/Divider with 10 Decoded Decimal Outputs

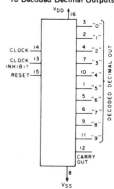

CD4018
Presettable Divide-by-"N" Counter Fixed or Programmable

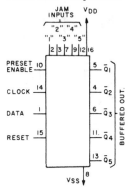

CD4019
Quad AND/OR Select Gate

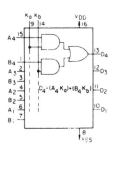

CD4020
14-Stage Binary Ripple Counter

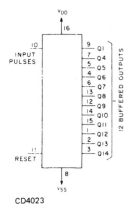

CD4021
8-Stage Static Shift Register Asynchronous Parallel or Synchronous Serial Input/ Serial Output

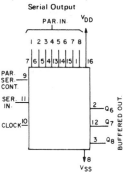

CD4022
Divide-by-8 Counter/Divider with 8 Decoded Decimal Outputs

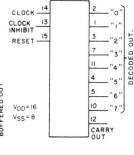

CD4023
Triple 3-Input NAND Gate

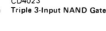

7-Stage Ripple-Carry Binary Counter/Divider

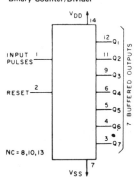

CD4025
Triple 3-Input NOR Gate

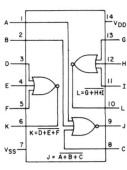

CD4026
Decade Counter/Divider with 7-Segment Display Outputs and Display Enable

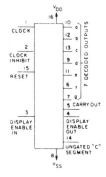

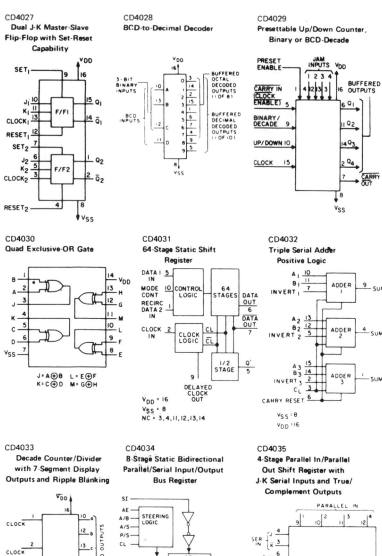

CD4027
Dual J-K Master-Slave Flip-Flop with Set-Reset Capability

CD4028
BCD-to-Decimal Decoder

CD4029
Presettable Up/Down Counter, Binary or BCD-Decade

CD4030
Quad Exclusive-OR Gate

$$J = A \oplus B \quad L = E \oplus F$$
$$K = C \oplus D \quad M = G \oplus H$$

CD4031
64-Stage Static Shift Register

$V_{DD} = 16$
$V_{SS} = 8$
NC = 3, 4, 11, 12, 13, 14

CD4032
Triple Serial Adder Positive Logic

$V_{SS} = 8$
$V_{DD} = 16$

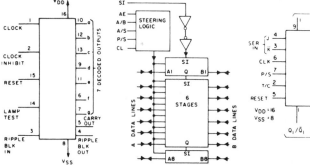

CD4033
Decade Counter/Divider with 7-Segment Display Outputs and Ripple Blanking

CD4034
8-Stage Static Bidirectional Parallel/Serial Input/Output Bus Register

CD4035
4-Stage Parallel In/Parallel Out Shift Register with J-K Serial Inputs and True/Complement Outputs

CD4037
Triple AND/OR Bi-Phase Pair

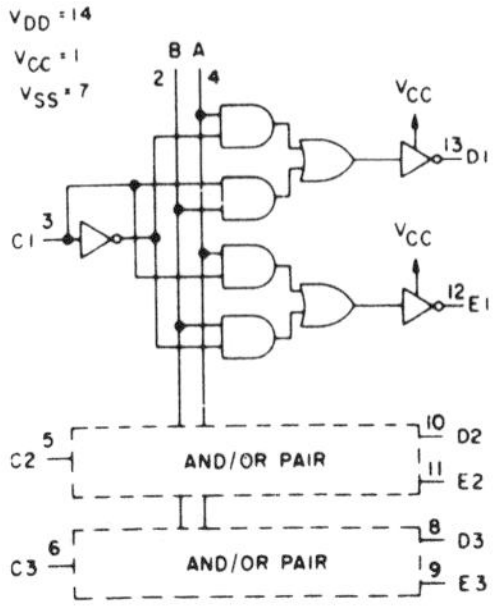

$V_{DD} = 14$
$V_{CC} = 1$
$V_{SS} = 7$

CD4038
Triple Serial Adder Negative Logic

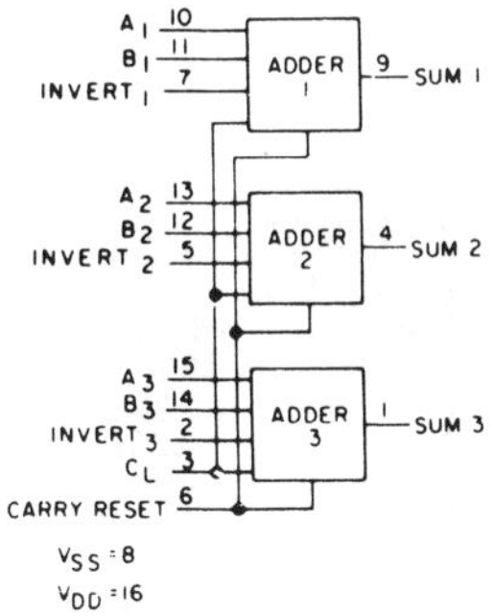

$V_{SS} = 8$
$V_{DD} = 16$

CD4040
12-Stage Ripple-Carry Binary Counter/Divider

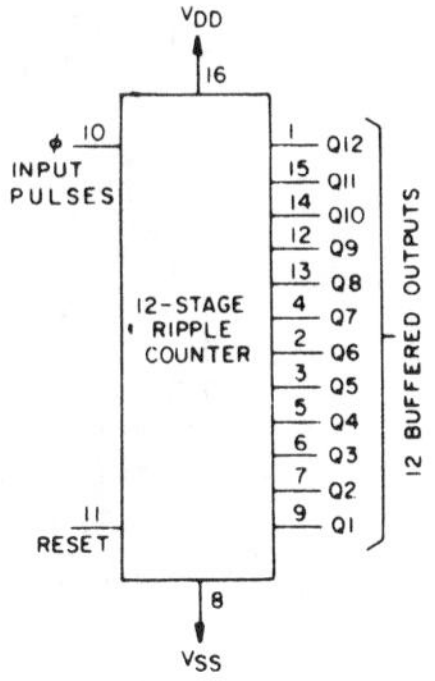

CD4041
Quad True/Complement Buffer

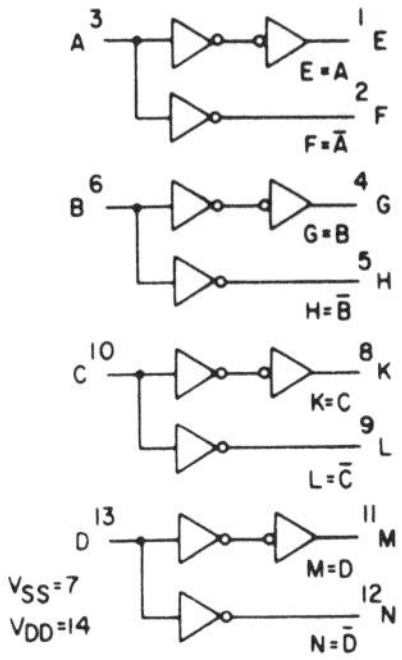

$V_{SS} = 7$
$V_{DD} = 14$

CD4042
Quad Clocked "D" Latch

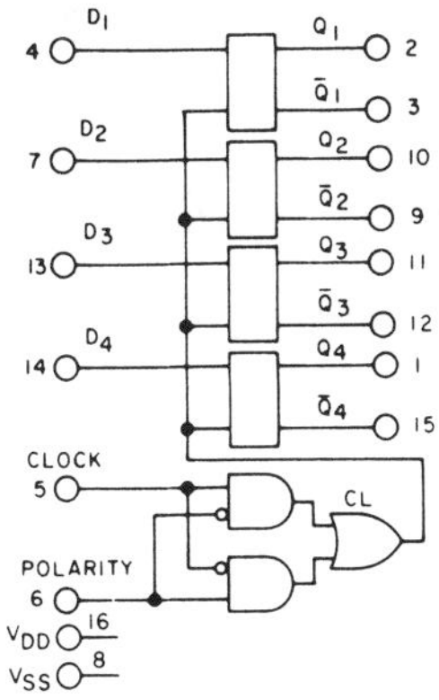

CD4043
Quad 3-State NOR R/S Latch

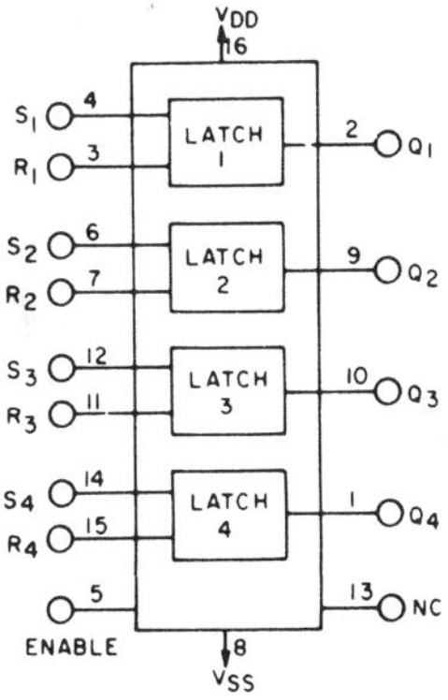

CD4044
Quad 3-State NAND R/S Latch

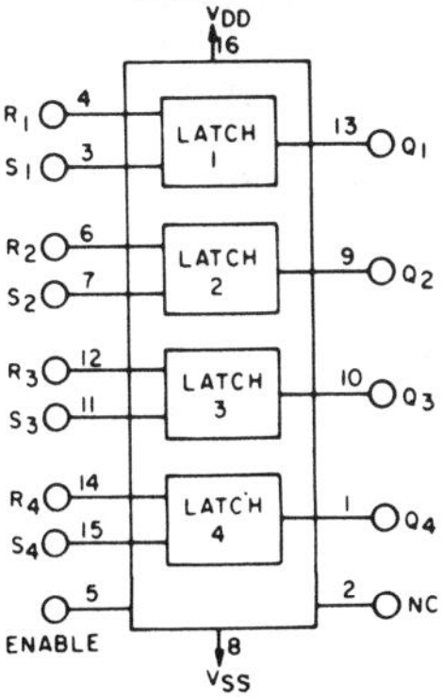

CD4045
21-Stage Counter

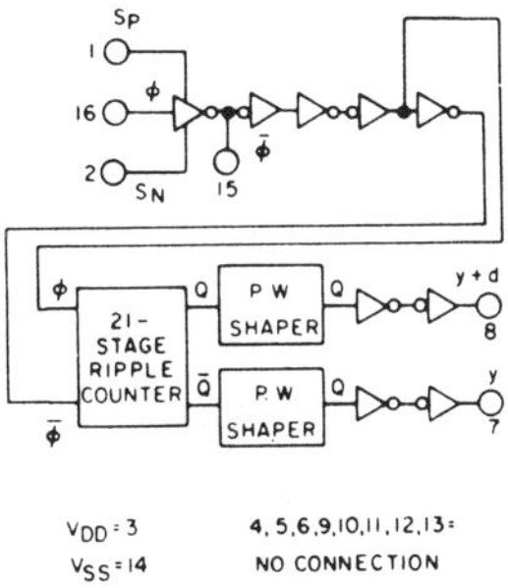

$V_{DD} = 3$
$V_{SS} = 14$

4,5,6,9,10,11,12,13 =
NO CONNECTION

CD4046
Micropower Phase-Locked Loop

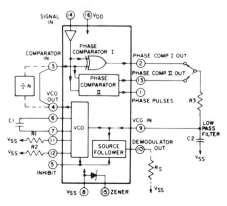

CD4047
Low-Power Monostable/Astable
Multivibrator

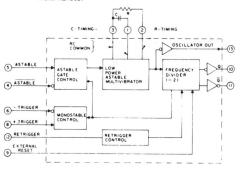

CD4048
Multi-Function Expandable
8-Input Gate

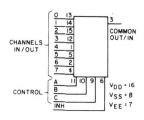

CD4049
Hex Buffer/Converter
Inverting Type

A $\frac{3}{}$ ▷ $\frac{2}{}$ $G = \overline{A}$

B $\frac{5}{}$ ▷ $\frac{4}{}$ $H = \overline{B}$

C $\frac{7}{}$ ▷ $\frac{6}{}$ $I = \overline{C}$

D $\frac{9}{}$ ▷ $\frac{10}{}$ $J = \overline{D}$

E $\frac{11}{}$ ▷ $\frac{12}{}$ $K = \overline{E}$

F $\frac{14}{}$ ▷ $\frac{15}{}$ $L = \overline{F}$

V_{CC} $\frac{1}{}$

V_{SS} $\frac{8}{}$

NC = 13
NC = 16

CD4050
Hex Buffer/Converter
Non-Inverting Type

A $\frac{3}{}$ ▷ $\frac{2}{}$ $G = A$

B $\frac{5}{}$ ▷ $\frac{4}{}$ $H = B$

C $\frac{7}{}$ ▷ $\frac{6}{}$ $I = C$

D $\frac{9}{}$ ▷ $\frac{10}{}$ $J = D$

E $\frac{11}{}$ ▷ $\frac{12}{}$ $K = E$

F $\frac{14}{}$ ▷ $\frac{15}{}$ $L = F$

V_{CC} $\frac{1}{}$

V_{SS} $\frac{8}{}$

NC = 13
NC = 16

CD4051
Single 8-Channel Analog
Multiplexer/Demultiplexer

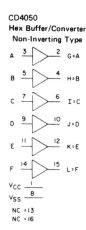

CD4052
Differential 4-Channel Analog Multiplexer/Demultiplexer

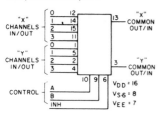

CD4053
Triple 2-Channel Multiplexer/Demultiplexer

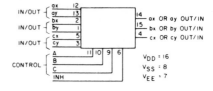

CD4054
4-Segment Liquid-Crystal Display Driver

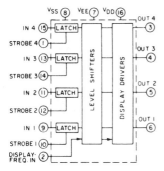

CD4055
BCD-to-7-Segment Decoder/Driver with "Display-Frequency" Output Liquid-Crystal Display Driver

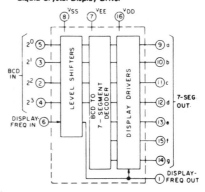

CD4056
BCD-to-7-Segment Decoder/Driver with Strobed-Latch Function Liquid-Crystal Display Driver

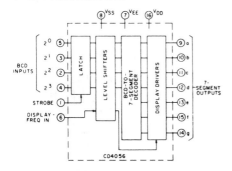

CD4057
4-Bit Arithmetic Logic Unit

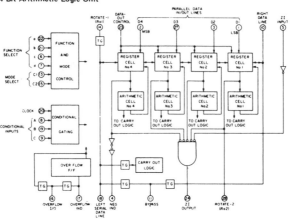

CD4059
Programmable Divide-by-"N" Counter

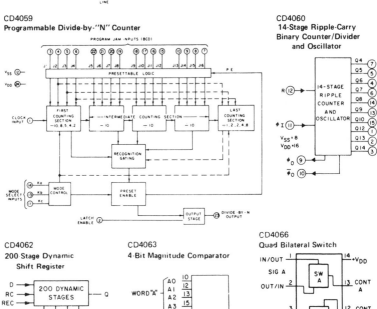

CD4060
14-Stage Ripple-Carry Binary Counter/Divider and Oscillator

CD4062
200 Stage Dynamic Shift Register

CD4063
4-Bit Magnitude Comparator

CD4066
Quad Bilateral Switch

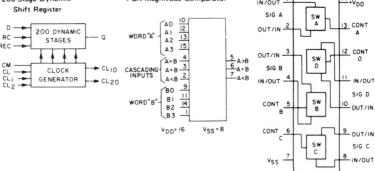

CD4067
16-Channel
Multiplexer/Demultiplexer

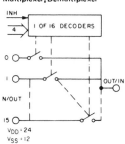

V_{DD} = 24
V_{SS} = 12

CD4068
8-Input NAND/AND Gate

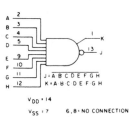

V_{DD} = 14

V_{SS} = 7   6, 8 = NO CONNECTION

CD4069
Hex Inverter

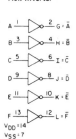

V_{DD} = 14
V_{SS} = 7

CD4070
Quad Exclusive-OR Gate

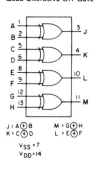

$J = A \oplus B$ $M = G \oplus H$
$K = C \oplus D$ $L = E \oplus F$

V_{SS} = 7
V_{DD} = 14

CD4071
Quad 2-Input OR Gate

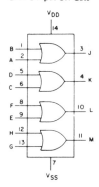

CD4072
Dual 4-Input OR Gate

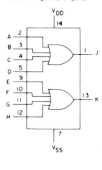

CD4073
Triple 3-Input AND Gate

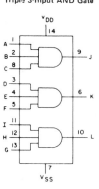

CD4075
Triple 3-Input OR Gate

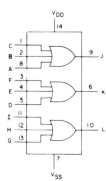

CD4076
4-Bit D-Type Register

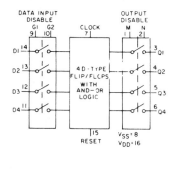

CD4077
Quad Exclusive-NOR Gate

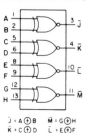

$\bar{J} = A \oplus B$ $\bar{M} = G \oplus H$
$\bar{K} = C \oplus D$ $\bar{L} = E \oplus F$

CD4078
8-Input NOR/OR Gate

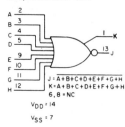

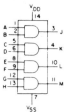

$J = \overline{A+B+C+D+E+F+G+H}$
$K = A+B+C+D+E+F+G+H$
$6, 8 = NC$

$V_{DD} = 14$
$V_{SS} = 7$

CD4081
Quad 2-Input AND Gate

CD4082
Dual 4-Input AND Gate

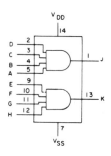

CD4085
Dual 2-Wide, 2-Input AND-OR-INVERT (AOI) Gate

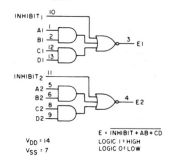

$E = \overline{INHIBIT + AB + CD}$
LOGIC 1 = HIGH
LOGIC 0 = LOW

$V_{DD} = 14$
$V_{SS} = 7$

CD4086
Expandable 4-Wide, 2-Input AND-OR-INVERT (AOI) Gate

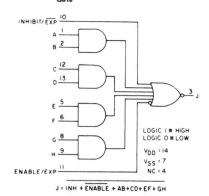

LOGIC 1 = HIGH
LOGIC 0 = LOW

$V_{DD} = 14$
$V_{SS} = 7$
$NC = 4$

$J = \overline{INH + \overline{ENABLE} + AB + CD + EF + GH}$

CD4089
Binary Rate Multiplier

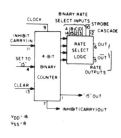

$V_{DD} = 16$
$V_{SS} = 8$

CD4093
Quad 2-Input NAND
Schmitt Trigger

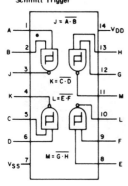

CD4094
8-Stage Shift-and-Store
Bus Register

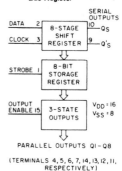

PARALLEL OUTPUTS QI—Q8

(TERMINALS 4, 5, 6, 7, 14, 13, 12, 11,
RESPECTIVELY)

CD4095
Gated J-K Master-Slave
Flip-Flop, Non-Inverting
Inputs

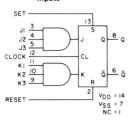

CD4096
Gated J-K Master-Slave
Flip-Flop, Inverting and
Non-Inverting Inputs

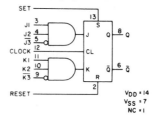

CD22100
4-by-4 Crosspoint Switch
with Control Memory

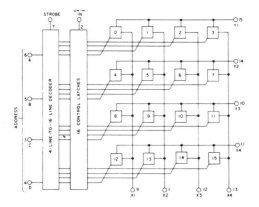

CD22101   CD22102
4-by-4-by-2 Crosspoint Switch with Control Memory

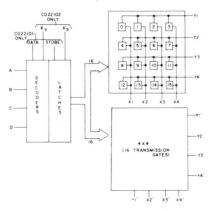

CD40100
32-Stage Static Left/Right Shift Register

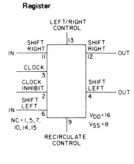

CD40101
9-Bit Parity Generator/Checker

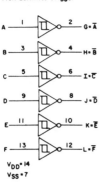

CD40102 2-Decade BCD
CD40103 8-Bit Binary
8-Stage Presettable Synchronous Down Counter

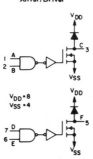

CD40105
FIFO Register 4-Bits Wide by 16-Bits Long

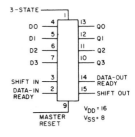

CD40106
Hex Schmitt Trigger

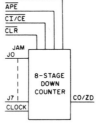

CD40107
Dual 2-Input NAND Buffer/Driver

CD40108
4-by-4 Multiport Register

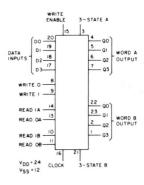

CD40109

**Quad Low-to-High
Voltage Level Shifter**

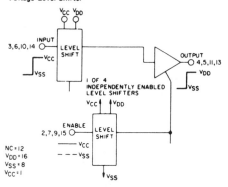

CD40110

**Decade Up-Down Counter/
Decoder/Latch/Driver**

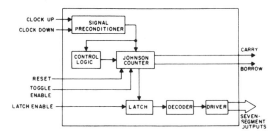

CD40114

64-Bit Random-Access Memory

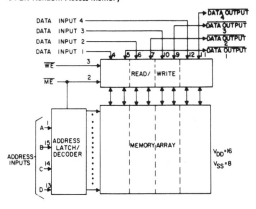

CD40115

8-Bit Bidirectional
CMOS/TTL Level Converter

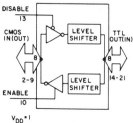

$V_{DD} = 1$
$V_{CC} = 22$
$V_{SS} = 11$

CD40147

10-Line-to-4-Line
BCD Priority Encoder

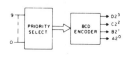

CD40160 Decade with Asynchronous Clear
CD40161 Binary with Asynchronous Clear
CD40162 Decade with Synchronous Clear
CD40163 Binary with Synchronous Clear

Synchronous 4-Bit Counter

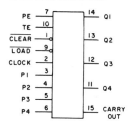

$V_{DD} = 16$
$V_{SS} = 8$

CD40174

Hex "D" Type Flip-Flop

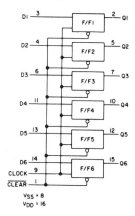

$V_{SS} = 8$
$V_{DD} = 16$

CD4097

Differential 8-Channel
Multiplexer/Demultiplexer

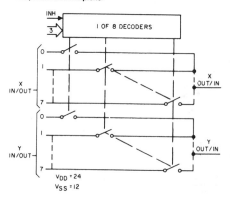

$V_{DD} = 24$
$V_{SS} = 12$

CD4098

Dual Monostable
Multivibrator

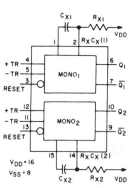

$V_{DD} = 16$
$V_{SS} = 8$

CD4099
8-Bit Addressable Latch

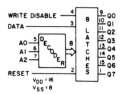

CD4502
Strobed Hex Inverter/Buffer

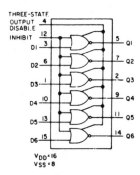

CD4508
Dual 4-Bit Latch

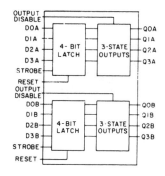

CD4510
BCD Presettable Up/Down Counter

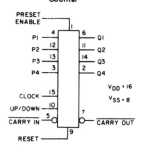

CD4511
BCD-to-7-Segment Latch Decoder Driver

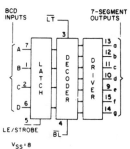

CD4512
8-Channel Data Selector

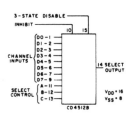

CD4514
CD4515 Output "Low" on Select
4-Bit Latch/4-to-16 Line Decoder

$V_{DD} = 24$
$V_{SS} = 12$

DATA 1 — 2 — A
DATA 2 — 3 — B — LATCH — 4 TO 16 DECODER
DATA 3 — 21 — C
DATA 4 — 22 — D
STROBE — 1

11 — S0
9 — S1
10 — S2
8 — S3
7 — S4
6 — S5
5 — S6
4 — S7
18 — S8
17 — S9
20 — S10
19 — S11
14 — S12
13 — S13
16 — S14
15 — S15

INHIBIT — 23

CD4516
Binary Presettable Up/Down Counter

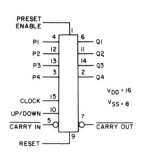

PRESET ENABLE
P1 — 4 — 6 — Q1
P2 — 12 — 11 — Q2
P3 — 13 — 14 — Q3
P4 — 3 — 2 — Q4
CLOCK — 15
UP/DOWN — 10 — 7
CARRY IN — 5
RESET — 9 — CARRY OUT

$V_{DD} = 16$
$V_{SS} = 8$

CD4517
Dual 64-Bit Shift Register

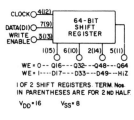

CLOCK O — 4(12)
DATA(DI) O — 7(9) — 64-BIT SHIFT REGISTER
WRITE ENABLE O — 3(13)
1(15) 6(10) 2(14) 5(11)

WE = 0 — Q16 — Q32 — Q48 — Q64
WE = 1 — DI7 — D33 — D49 — HiZ

1 OF 2 SHIFT REGISTERS. TERM. Nos.
IN PARENTHESES ARE FOR 2 ND HALF.

$V_{DD} = 16$ $V_{SS} = 8$

CD4518 BCD
CD4520 Binary
Dual Up Counter

CLOCK A — 1 — C — ÷10/-16 — 3 Q1A / 4 Q2A / 5 Q3A / 6 Q4A
ENABLE A — 2
RESET A — 7 — R

CLOCK B — 9 — C — -10/-16 — 11 Q1B / 12 Q2B / 13 Q3B / 14 Q4B
ENABLE B — 10
RESET B — 15 — R

$V_{DD} = 16$
$V_{SS} = 8$

CD4527
BCD Rate Multiplier

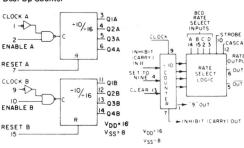

BCD RATE SELECT INPUTS
A B C D — 14 15 2 3
CLOCK — 9
INHIBIT (CARRY) IN — 11
SET TO NINE — 4 — ÷10 COUNTER
CLEAR — 13
STROBE — 10
CASCA — 12
RATE SELECT LOGIC — RATE OUTPUT / 6 OUT / 5 OUT
9 OUT — 1
7 — INHIBIT (CARRY) OUT

$V_{DD} = 16$
$V_{SS} = 8$

CD4532
8-Bit Priority Encoder

D7 — 4
D6 — 3
D5 — 2
D4 — 1 — PRIORITY SELECT
D3 — 13
D2 — 12
D1 — 11
D0 — 10

ENCODER — 6 Q2 / 7 Q1 / 9 Q0

15 E_O
14 GS
E_i — 5

$V_{DD} = 16$
$V_{SS} = 8$

CD4536
Programmable Timer

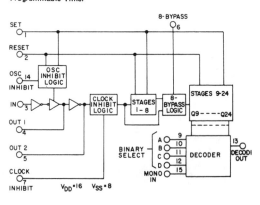

SET O — 1
RESET O — 2
OSC INHIBIT O — 14 — OSC INHIBIT LOGIC
IN O — 3
OUT 1 O — 4
OUT 2 O — 5
CLOCK O — 7
INHIBIT

CLOCK INHIBIT LOGIC
STAGES 1-8
8-BYPASS LOGIC
STAGES 9-24 — Q9 — Q24

8-BYPASS O — 6

BINARY SELECT — A O — 9 / B O — 10 / C O — 11 / D O — 12
MONO IN O — 15

DECODER — 13 DECODING OUT

$V_{DD} = 16$ $V_{SS} = 8$

CD4555
**Dual Binary-to-1-of-4
Decoder/Demultiplexer
Output "High" on Select**

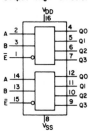

CD4556I
**Dual Binary-to-1-of-4
Decoder/Demultiplexer
Output "Low" on Select**

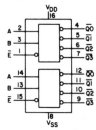

CD4585
4-Bit Magnitude Comparator

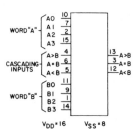

CD4724
8-Bit Addressable Latch

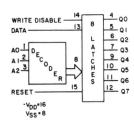

CD40181
4-Bit Arithmetic Logic Unit

Active-Low Data · Active-High Data

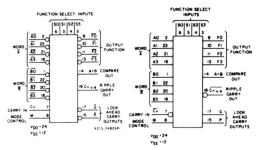

CD40182
Look-Ahead Carry Generator

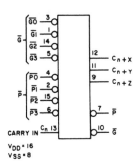

CD40192 BCD
CD40193 Binary
**Presettable Up/Down Counter
(Dual Clock with Reset)**

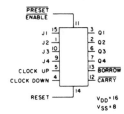

CD40208
4-by-4 Multiport Register

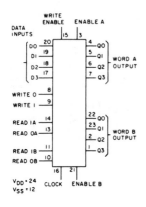

CD40257

**Quad 2-Line-to-1-Line
Data Selector/Multiplexer**

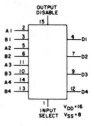

Chapter 6

Microprocessors and Microprocessor Systems

A microprocessor is a programmable logic chip which can make use of memory. The microprocessor can address memory, which means that it can select stored data and make use of it and store results also in memory. Within the microprocessor chip itself, logic actions such as the standard NOT, AND, OR and XOR gate actions can be carried out on a set of bits, as well as a range of other register actions such as shift and rotate, and some simple arithmetic. The fact that any sequence of such actions can be carried out under the control of a program which is also read from memory is the final item that completes the definition of a microprocessor.

In general, microprocessors are designed so as to fall into one of two classes. One type is intended for industrial control, and this also extends to the control of domestic equipment, such as central heating systems. A microprocessor of this type will often be almost completely self-contained, with its own memory built in, and very often this will include the programming instructions. Such microprocessors will very often need to work with a limited number of bits at a time, perhaps 4. The number of possible programming instructions need only be small. The control microprocessor will also be offered typically as a 'semi-custom' device, with the programming instructions put in at the time of manufacture for one particular customer. Some microprocessors originally intended for computing purposes, such as the 6502 and the Z80, are now also being used in controllers.

The alternative is the type of microprocessor whose main purpose is computing. These are of two types: the complex instruction set (CISC) and the reduced instruction set (RISC) types. Either type can be manufactured with little or no memory of its own, but each is capable of addressing large amounts of external memory and will deal with at least 16 bits, and more usually 32 bits of data at a time. The CISC type has a much larger range of instructions, and will generally operate at fairly high speeds. The RISC type uses very few instructions, each of which can be executed very rapidly. The thinking behind RISC is that most microprocessors spend most of their working lives carrying out a relatively small number of instructions, so that by concentrating on the fast and efficient processing of these few instructions the processor will be faster. In practice, the need for more complicated actions to be carried out requires software for the RISC type of processor to be longer and more elaborate, reducing its advantages. Several types of computing processors now contain appreciable amounts of memory (64K typically) used as temporary storage (cache memory). This memory is fast (25 ns or less) and is used to overcome the problem of slow main memory by reading or writing the main memory at a time when the microprocessor is otherwise occupied.

The instruction register

Each microprocessor chip contains one main control register which is designated as the instruction register. This will be of as many bits as the microprocessor is designed to handle at a time, usually 16 or 32. Whatever is put into this instruction register completely decides what the microprocessor will do, so that access to this register must be very severely limited. Most of the bytes that it deals with, in fact, come from a preset group that is permanently stored within the microprocessor, called the microprogram. By using this system, the makers of microprocessors avoid the need to have to check each input to the microprocessor in case it should contain conflicting commands, such as connecting all registers together. The principle is that a set of microprograms is stored in fixed memory inside the microprocessor. There will be one program for adding, another for subtracting, one for ANDing, another for ORing and so on. Putting a set of bits into the instruction register will have the effect of calling up one of these microprograms. If the bits in the instruction register correspond to the code number for a microprogram, then the microprogram is run. This is done by feeding groups of microprogram bits into the instruction register in turn until the process is complete. If the group of bits that is used to call up the microprogram does not correspond to any existing microprogram, then the command is ignored. In this way, the gating within the microprocessor is controlled in a completely predictable way, one that has been determined by the manufacturers and is built into the chip. Though the system is not totally infallible – most microprocessors have 'undocumented' instructions due to unintentional inclusion of codes – this system of using a restricted menu of instructions serves well, and for the RISC type of processor the number of microprograms can be very small.

The loading of a command into the instruction register is dealt with by part of the microprogramming. When the microprocessor is switched on, its first action will be to load in a word (16 bits) or double-word (32 bits, a Dword), according to whether it is a 16-bit or a 32-bit device – older microprocessors worked with single-byte 8-bit units. This first word or Dword is gated directly into the instruction register, and then these gates are disabled. If this word or Dword now calls up a microprogram, the instruction register is supplied from the internal microprogram, and gates will allow signals to be copied from one register to another or allow external signals from the pins of the chip to be connected to other internal registers following the instructions of the microprogram. Following the last microprogram action, the gates which connect from the pins into the instruction register are enabled again so that another command word or Dword can be read. In this way, the microprocessor reads commands only when it is ready for them. Since each action is triggered by the arrival of a clock pulse, the timing is always exact, and each instruction will take a fixed (but different) number of clock pulses to run its microprogram.

Take, for example, the process of adding two numbers which we will imagine are already stored in two registers, with the result to be returned to one of the registers. The action of addition is started when the microprocessor reads an ADD instruction word or Dword taken from the external memory. The ADD instruction is gated directly into the instruction register, which is then shut off as far as external signals are concerned. By analysing the bits of the instruction, the correct microprogram is called up. The first part of the microprogram is then loaded into the instruction register, and its action is to connect one register to one set of the inputs to an adder, a collection of gates, with two sets of inputs and one set of outputs. This action requires one clock pulse, and on the next clock pulse the other register is connected to the other set of adder inputs. The next clock pulse provides the next microprogram instruction, which connects the (stored) output of the adder back to the input of one register. The next clock pulse brings in the microprogram action which enables the gates, so adding the bits and then storing the result back on the next clock pulse. The last microprogram action must then re-enable the gates which allow a command to enter the instruction register from outside the microprocessor.

This is oversimplified, particularly as regards the clock pulses, which are also gated to the correct places. The principle, however, is that the actions in the ADD routine are decided by the microprogram which has been called into action by a single instruction. In addition, we have established the very important point that this single instruction reaches the instruction register only when gates are enabled, and that for most of the time the instruction register is not accessible to signals from outside the microprocessor. Another important principle is that all actions are carried out in sequence, one stage of action for each step in the microprogram. Finally, all of the actions are controlled by the clock pulses, and the speed of all processing depends on the speed of these clock signals.

The construction of many types of microprocessor chips makes it impossible to observe the action with slow clock pulses. This is because signal voltages leak from one point to another within the microprocessor, and it is only at high clock speeds that the time for leakage becomes so short that the voltage levels are not significantly affected. As a rough rule of thumb, microprocessors should not be run at a clock speed much below 100 kHz, and some cannot be run even at this rate. This does not apply to some CMOS types, such as the Intel HCMOS 80C86, which can be run at clock speeds down to zero. This allows the device to be clocked, if needed, by a push-button, so that the actions can be analysed in detail. The maximum clock speed is determined by the design of the micro-processor, particularly the stray capacitances. For modern microproces-sors, maximum clock speeds in the range of 12 MHz to 33 MHz are common. Since the clock rate decides the rate of processing, micro-processors used in computers are usually run as fast as the chip can reliably cope with, and clock speeds up to 50 MHz are currently in use.

Clocking

The clock pulse is the master timing pulse of any microprocessor system so that the specification of the clock pulse in terms of repetition rate, rise and fall times and pulse shape is fairly exacting. Some microprocessor chips include their own oscillator circuits, so that the only external components that need to be connected are a quartz crystal of the correct resonant frequency, and a few other discrete components. More often, the clock is an external circuit which can be a chip supplied by the makers of the microprocessor, or a circuit constructed from logic gates of the TTL LS class. *Figure 6.1* shows typical clock specifications for older Intel microprocessors working with clock periods of 200 to 500 ns. The rise and fall times are particularly important, because slow rise and fall times can cause considerable timing problems in circuits that use microproces-sors.

The specification of a minimum clock frequency, which can be as high as 8 MHz for a microprocessor that normally operates with a 12 MHz clock, reflects the high leakage inside the chip, due to very close packing of tracks. The requirement for maintaining fast rise and fall times for the clock pulse means that some care has to be taken with circuitry around the clock terminals of the microprocessor. This is not difficult if the clock circuits are built into the chip, or even if an external clock chip is used, because this can be located near the clock input pin of the microproces-sor. Where problems may be encountered is when a co-processor is used, which is another microprocessor running along with the main micro-processor. This co-processor will need to be supplied with the same clock

All timings in nanoseconds, ns.

Symbol	Timed period	8088 Min	Max	80C88A Min	Max	80C88AL2 Min	Max	Notes
TCLCL	CLK cycle period	200	500	125	DC	125	DC	
TCLCH	CLK low time	118		68		68		
TCHCL	CLK high time	69		44		44		
TCH1CH2	CLK rise time		10		10		10	1
TCL2CL2	CLK fall time		10		10		10	2

Note 1: From 1.0 V to 3.5 V levels
Note 2: From 3.5 V to 1.0 V levels

Figure 6.1. Typical clock pulse specifications, in this example for the Intel 80xx range of microprocessors

pulse, usually in phase, and with the same requirements of rise and fall times. Unless this co-processor can be located on the same board and very close to the main processor, the stray capacitance of PCB tracks along with the load capacitance of the clock input pins of the co-processor can cause degradation of the pulse shape. The problem is very much greater if the co-processor is located on another board. In this case, the use of Schmitt trigger stages both at the output from the main clock and the input to the co-processor clock will usually be needed to maintain fast rise and fall times.

Memory

Memory is a name for a type of digital circuit component which can be of several different types. The basis of a single unit of memory is that it should retain a 0 or a 1 signal, and that this signal can be connected to external lines when needed for reading (copying) or writing (storing). The method that is used to enable connection is called addressing. The principle is that each unit of memory should be enabled with a unique combination of signals that is present on a set of lines, called the address lines or the address bus. Since each combination of bits constitutes a binary number, the combinations are called address numbers and are usually expressed in hexadecimal because of their unwieldy size in binary.

Bearing in mind that memory consists of stores of binary digits that are connected to external pins by using an address number, then we can examine the methods that are used to implement this principle. To start with, there are two fundamental types of memory, both of which are needed in virtually any microprocessor application. One type of memory contains fixed bits, unalterable, and is therefore called 'read-only' memory, or ROM. The important features of ROM is that it is non-volatile, meaning that the stored bits are unaffected by switching off power to the memory and are available for use whenever power is restored. Since there must be an input to the microprocessor whenever it is switched on, ROM is essential to any microprocessor application, and in some applications it might be the only type of memory that is needed. The simplest type of such a ROM consists of permanent connections to logic 0 or logic 1 voltage lines, gated to output pins when the correct address number is applied to the address pins. A typical ROM might use eight sets of demultiplexers and gates, each using 16 address lines and with 65 536 gates to each output pin. In other words, this is a 64K × 8-bit ROM. This type of ROM is called a 'masked' ROM, referring to the i.c. manufacturing technique in which etching masks determine the layout of connections. The masks form the main initial cost of production of such a ROM, and the use of masked ROM is feasible only if the content of this memory is thoroughly tested and proven.

An alternative to masked ROM is some form of EPROM, the Erasable Programmable Read-only Memory. There are several varieties, but nearly all use the same principles – of making connections through lightly doped semiconductor by injecting carriers (electrons or holes), which are then trapped. The system of gating and demultiplexing which forms the addressing for the chips is the same, and only the method of connecting to logic 1 or 0 through paths in the semiconductor is different. The point about a PROM is that the connections can be established by connecting to the inputs of gates which are then 'blown' by using higher than normal voltages (typically 16 to 25 V for a chip that normally operates at 5 V) in a programming cycle. This normally consists of cycling several times through each address number, applying the high voltage for each logic 1 bit that is needed, with the correct signals taken from a temporary memory source. Once programmed or 'blown' in this way, the PROM can be used like a ROM. The advantage is that a PROM is manufactured blank, there is no special masking cost, and, more importantly, no extra design time is needed in the manufacturing process. If the programming is faulty, new PROMS can be blown and tried, until the system seems to be bug free. If this is done at prototype stage, the result can be a very reliable piece of equipment, and a masked ROM can be made from any copy of the PROM. Using PROMs is a short-term expedient, but one which is useful in that short term.

Though EPROM chips are expensive, they can be re-used, unlike the older 'fusible link' type in which the internal connections are opened permanently during the blowing process. The most popular type of EPROM is erasable by shining UV light into the silicon whose conductivity establishes the logic 1 connections. The effect of UV is to make the material momentarily conductive to such an extent that the trapped charges can move away, making the material into an insulator when the UV no longer strikes it. This process is described as 'PROM-washing', and typically takes an exposure of 5 minutes to 30 minutes, depending on the construction of the PROM and the wavelength of UV. The most effective UV for the purpose is the shorter-wavelength type, and this radiation must not be allowed to reach any part of a human body, particularly the eyes. Prom-washers must therefore be constructed in light-tight boxes, with interlock switches to eliminate the possibility of the light being on when the box is open.

Read–write memory

Read–write memory is the other form of memory which for historical reasons is always known as RAM (Random-access Memory). This is because, in the early days, the easiest type of read–write memory to manufacture consisted of a set of serial registers, from which bits could be read at each clock pulse. The construction of a memory from which a bit could be selected by using an address number was a much more difficult task. The use of addressing means that any bit can be selected at random, without having to feed out all the preceding bits; hence the name 'random access'. Practically all forms of memory that are used nowadays in microprocessor systems feature random-access addressing, but the name has stuck as a term for read–write memory. Memory of this type needs address pins, data pins and control pins to determine when the chip is enabled and whether it is to be written or read. The reading of a memory of this type does not alter the contents.

Unlike ROM, RAM is normally volatile. It is possible, however, to fabricate memory using CMOS techniques and retain the data for very long periods, particularly if a low-voltage backup battery can be used. Such CMOS RAM is used extensively in calculators and is used in computers to hold essential set-up data, including time and date information. Static RAM is based on a flip-flop as each storage bit element. The state of a flip-flop can remain unaltered until it is deliberately changed, or until power is switched off, and this made static RAM the first choice for manufacturers in the early days. The snag is that power consumption can be large, because each flip-flop will draw current whether it stores a 0 or a 1. This has led to static RAM, except for the CMOS variety, being used only for comparatively small memory sizes and where fast operation is needed.

The predominant type of RAM technology for large memory sizes is the dynamic RAM. Each cell in this type of RAM consists of a miniature MOS capacitor with logic 0 represented by a discharged capacitor and logic 1 by a charged capacitor. Since the element can be very small, it is possible to construct very large RAM memory chips ($1M \times 1$-bit, $256K \times 4$-bit are now common), and the power requirements of the capacitor are very small. The snag is that a small MOS capacitor will not retain charge for much longer than a millisecond, since the connections to the capacitor will inevitably leak. All dynamic memory chips must therefore be 'refreshed', meaning that each address which contains a logic 1 must be re-charged at intervals of no more than a millisecond. The refreshing action can be carried out within the chip, providing that the cycling of address numbers is done externally, and is 'user-transparent', meaning that the user is never aware of the action.

The buses

Bus is the name that is given to a set of lines in a microprocessor circuit. The three main buses are the address bus, the data bus, and the control bus. The buses of a microprocessor system consist of lines that are connected to each and every part of the system, so that signals are made available at many chips simultaneously. Since understanding the bus action is vitally important to understanding the action of any micro-processor system, we will concentrate on each bus in turn, starting with the address bus.

An address bus consists of the lines that connect between the microprocessor address pins and each of the memory chips in the microprocessor system. In anything but a very simple system, the address bus would connect to other units also, but for the moment we will ignore these other connections. A typical older-style 8-bit microprocessor would use 16 address pins. Using the relationship that n pins allow 2^n binary number combinations, the use of 16 address lines permits 65 536 memory addresses to be used, and modern computing microprocessors use 20, 24 or 32 address lines. Many types of memory chips are '1-bit' types, which allow only 1 bit of data to be stored per address. For a 64K 8-bit microprocessor, then, the simplest RAM layout would consist of eight 64K $\times$ 1-bit chips, each of which would be connected to all 16 lines of the address bus. Each of these chips would then contribute 1 bit of data, so that each chip is connected to a different line of the data bus. This scheme is illustrated in *Figure 6.2*. At each of the 65 536 possible address numbers, each chip will give access to 1 bit, and this access is provided through the lines of the data bus. The combination of address bus and data bus provides for addressing and the flow of data, but another line is needed to determine the direction of data.

This extra line is the read/write line, one of the lines of the control bus. When the read/write line is at one logic level, the signal at each memory chip transfers all connections to the inputs of the memory units, so that the memory is written with whatever bits are present on the data lines. If the read/write signal changes to the opposite logic level, then the internal gating in the memory chips connects to the output of each memory cell rather than to the input, making the logic level of the cell affect the data line. The provision of address bus, data bus and read/write line will therefore be sufficient to allow the microprocessor to work with 64K of memory in this example. For smaller amounts of memory, the only change

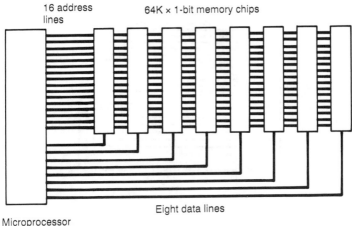

Figure 6.2. The arrangement of address lines and data lines for a simple 64K × 1-bit memory. Each memory chip is connected to all of the address lines and to one data line

to this scheme is that some of the address lines of the address bus are not used. These unused lines must be the higher-order lines, starting at the most significant line. For a 16-line address bus, the most significant line is designated as A15, the least significant as A0.

A memory system for an 8-bit processor that consisted purely of 64K of RAM, however, would not be useful, because no program would be present at switch-on to operate the microprocessor. There must be some ROM present, even if it is a comparatively small quantity. For some control applications, the whole of the programming might use only ROM, and the system would consist of one ROM chip connected to all of the data bus lines, and as many of the address lines as were needed to address the chip fully. As an example, *Figure 6.3* shows what would be needed in this case, using an 8K × 8-bit ROM, which needs only the bottom 13 address lines. It is more realistic to assume that a system will need both ROM and RAM, and we now have to look at how these different sets of memory can be addressed. In the early days, the total addressing capability of an 8-bit machine was no particular restriction, and a common configuration was of 16K ROM and 16K RAM. This could be achieved by 'mapping' the memory as shown in *Figure 6.4* – other combinations are, of course, possible. In the scheme that is illustrated, the ROM uses the first 16K of address, and the RAM uses the next 16K. Now the important thing about this scheme is that 16K corresponds to 14 lines of an address bus, and the same 14 lines are used for both sets of memory. The principle is illustrated in *Figure 6.5*. The lower 14 address lines, A0 to A13, are connected to both sets of chips, represented here by single blocks. Line A14, however,

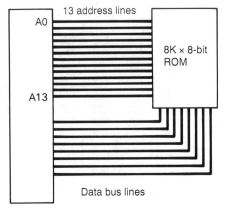

Figure 6.3. An example of connections between a microprocessor and an 8K × 8-bit ROM chip. Only 13 address lines are needed, labelled as A0 to A13

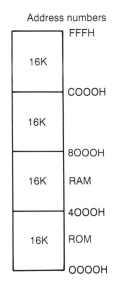

Figure 6.4. A typical memory map for the address area of an 8-bit microprocessor using small amounts of ROM and RAM

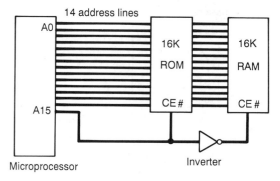

Figure 6.5. Using chip-enable signals to switch between ROM and RAM chips that use the same set of lower address lines

is connected to 'chip-enable' pins, which as the name suggests enable or disable the chips. The principle is simple. During the first 16K of addresses, line A14 is low, so that RAM is enabled (imagining the enable pin as being active when low) and ROM is disabled. For the next 16K of addresses on lines A0 to A13, A14 is high, so that RAM is disabled and ROM is enabled. This allows the same 14 address lines to carry out the addressing of both ROM and RAM. A simple scheme like this is possible only when both ROM and RAM occupy the same amount of memory and require the same number of address lines.

A very common scheme that was once used for small computers and is still found on industrial controllers is illustrated in *Figure 6.6*. The RAM consists of eight chips of 64K × 1-bit dynamic RAM, using the whole address range. The problem of how to deal with the ROM can then be solved in one of two ways. One is simply to map the ROM over some of the RAM. This means that a range of addresses will select ROM rather than RAM, and the RAM which exists in this range of addresses is never used. It looks wasteful, but the falling prices of 64K RAM chips have, in fact, made a scheme of this type cheaper than a memory built up from 16K chips and with no redundant blocks. The ROM is selected by using OR gating, and the diagram illustrates the situation in which 16K at the top end of memory is used for ROM in this way.

An alternative is to use 'bank-switching' schemes in which sets of 64K of memory are switched in or out as required. Modern computers require memory sizes which are much larger than bank-switching can easily provide, and the use of 24 or 32 address pins allows large amounts of memory to be addressed without resort to any switching tricks. This does not necessarily make these chips easier to work with, however. One of the problems for designers is the chip package itself. If the conventional 40-pin inline package is retained, it is not possible to cater for 20 address

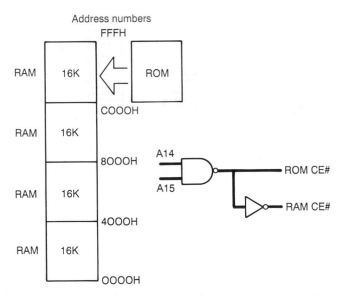

Figure 6.6. Mapping a ROM set of addresses over existing RAM addresses. This prevents some of the RAM from being used, but is often less costly than alternatives. The gating for the chip-enable pins is illustrated

lines and, at the same time, for more than eight data lines. The solution has been to multiplex pin use. This results in considerable hardware complications, and reduces the processing speed to a smaller fraction of the clock rate. This can be compensated for by designing the chip so that it can use higher clock rates. We can take as an example of this technique one of the most successful of the true 16-bit chips in a 40-pin package, the Intel 8086.

The pinout is illustrated in *Figure 6.7*, and you can see that all of the address pins serve a dual purpose. The pins AD0 to AD15 carry both address and data signals, and the pins A16/S3 to A19/S6 are used both for higher memory address bits and also for bus status or bank-switching signals. These pins are shared in a time-switching basis, using the clock pulses in sets of four. On the first clock pulse, all of the 20 address lines hold address data, which can be latched to the address bus. Data is transferred on subsequent clock pulses, depending on whether data is being read or written. Reading takes place on the third clock pulse, allowing time for the address bus to settle. Writing can start on the second clock pulse, after the write control signal has switched to logic 0. A complete group of four clock pulses is known as a bus cycle, and each clock pulse is referred to as a 'T-state'. *Figure 6.8* illustrates a simplified

Pin configuration – the # sign indicates active LOW. Minimum configuration allocations are shown first, maximum (where applicable) second.

1. GND	40. V$_{cc}$	
2. A14	39. A15	
3. A13	38. A16/S3	
4. A12	37. A17/S4	
5. A11	36. A18/S5	
6. A10	35. A19/S6	
7. A9	34. SSO#	High	
8. A8	33. MN/MX#	
9. AD7	32. RD#	
10. AD6	31. HOLD	RQ#/GT0#
11. AD5	30. HLDA	RQ#/GT1#
12. AD4	29. WR#	LOCK#
13. AD3	28. IO/M#	S2#
14. AD2	27. DT/R#	S1#
15. AD1	26. DEN#	S0#
16. AD0	25. ALE	QS0
17. NMI	24. INTA#	QS1
18. INTR	23. TEST#	
19. CLK	22. READY	
20. GND	21. RESET	

Figure 6.7. The pin allocation for the Intel 8086 16-bit microprocessor, showing the sharing of some data and address lines

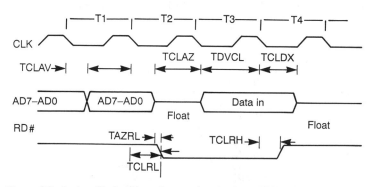

Figure 6.8. A simplified timing diagram for the Intel 8086, showing how the multiplexing of the pins is dealt with

version of the timing diagram, in which only the address and data line signals are shown.

The alternative to multiplexing the pin outputs is to use a larger package. The Motorola 68010 chip uses either a 64-pin dual-in-line package, or the more modern leadless square chip carrier package or pin-grid array, with 68 pins. The Intel 80286 also uses the JEDEC chip carrier or the 68-pin-grid array (PGA). The form of one of these grid arrays

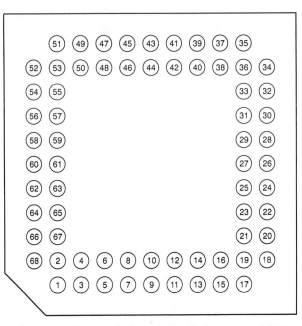

The pin assignment table uses the hash (#) to indicate active LOW.

1. BHE #
2. NC
3. NC
4. S1 #
5. S0 #
6. PEACK #
7. A23
8. A22
9. V_{ss}
10. A21
11. A20
12. A19
13. A18
14. A17
15. A16
16. A15
17. A14
18. A13
19. A12
20. A11
21. A10
22. A9
23. A8

24. A7
25. A6
26. A5
27. A4
28. A3
29. RESET
30. V_{cc}
31. CLK
32. A2
33. A1
34. A0
35. V_{ss}
36. D0
37. D8
38. D1
39. D9
40. D2
41. D10
42. D3
43. D11
44. D4
45. D12
46. D5

47. D13
48. D6
49. D14
50. D7
51. D15
52. CAP
53. ERROR #
54. BUSY #
55. NC
56. NC
57. INTR
58. NC
59. NMI
60. V_{ss}
61. PEREQ
62. V_{cc}
63. READY #
64. HOLD
65. HLDA
66. COD/INTA #
67. M/IO #
68. LOCK #

Figure 6.9. A pinout for the 68-pin-grid array used for some modern microprocessor chips – 132-pin grids are also used

is shown in *Figure 6.9* – this is the Intel version which has one corner cut to indicate the position of pin 1. The pins are in two rows around the square carrier, which measures approximately 28.7 mm square, just over a square inch. The use of this package demands a very good-quality PCB construction, and some care about handling if pins are being checked or tested with probes. The use of these packages, however, avoids multiplexing and allows the buses to be connected in more straightforward ways.

Reading and writing actions

Reading, as applied to a microprocessor, means that the data signals on the data bus are copied into a register within the microprocessor. As we have seen earlier, the first of a set of such data signals will always be copied into the instruction register. When the microprocessor reads a memory location, the bits of that memory location are completely unaffected, and the only changes take place within the microprocessor, with a register in the microprocessor being set into the same bit arrangement. The action of writing is another copying action which means that the bits in a register of the microprocessor are made available on the data bus, and used to alter either memory or another external register. When writing is to memory, the memory content will be changed, but the content of the register of the microprocessor which provides the signal is not.

Both reading and writing are actions which cause change, either in a microprocessor register or in the memory, and change takes time. To read from or write to memory means that a set of memory cells must be selected. In a 128K × 1-bit memory, for example, 1 bit in each chip must be selected. This is done, as we have seen, by the signals on the address bus, but some time will inevitably elapse between the signals on the bus becoming steady and the memory selection becoming complete. This time is a quoted parameter for each type of memory chip, and a typical figure for the dynamic memory fitted in small computers is of the order of 80 to 100 ns. By contrast, static RAM can provide times of 25 ns or lower. To this we need to add the time between getting the address bits into a register in the microprocessor, and the voltages on the lines becoming steady. Typical times for this operation are 10 to 100 ns. If a modest 4 MHz clock rate is being used, then the clock voltage will be high for around 120 ns, and low for another 120 ns, so that you cannot expect a memory to be read from or written to in the same clock cycle as started the addressing process. It would obviously be disastrous to attempt to read from or write to memory until the memory selection was complete, so that the timing of microprocessor operations is of vital importance. This timing is provided for in the design of the microprocessor, and this is a topic that we must now look at in more detail.

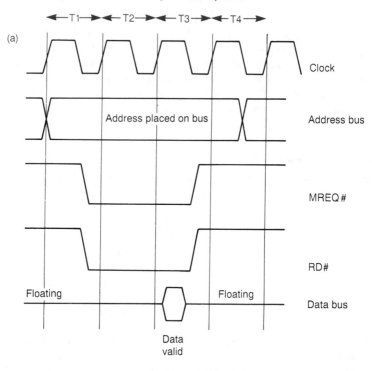

Figure 6.10. The data byte read cycle for the Z80. Note the use of the # sign to indicate an active-low signal. The time needed to transfer the data is very short

To avoid the complications of multiplexing and the extra facilities of the later types of microprocessors, we shall look at the timing for the well-established Z80, a chip that is still used in huge numbers. The Z80 uses timing cycles that differ (in control signals) according to whether an instruction code is being read, a byte of data is being read, or a byte of data is being written. Since most microprocessors use the same type of timing for reading any type of byte, with variations of control signals only, we will concentrate on the ordinary data byte read of the Z80, which is illustrated in *Figure 6.10*. The complete action requires three clock cycles to be completed, and the important point to appreciate is the relative timing of each part of the action. The read cycle starts at the leading edge of the first clock cycle. A short time after this leading edge, the memory address appears on the address bus. The delay is due to the time needed for the switching actions. The address is held in a 16-bit register called the program counter (PC), and at the leading edge of the clock pulse, the gates which connect each unit of the PC to the address pins are enabled. The presence of address bits is indicated on the timing diagram by the two parallel lines, one at each logic level, and the point where these cross over is where the address bits change. This address remains on the address bus throughout all three clock cycles, and does not change until there is a subsequent read/write action. The reading action, however, does not take place at this point because time is needed for the memory chips to respond. At the trailing edge of this first clock pulse, the MREQ (Memory Request) signal voltage goes low. The MREQ signal is one of the control bus signals, and its use allows the microprocessor to work with either memory or port (input/output) signals. At the same time, the RD (Read) signal voltage goes low to enable the memory chips.

The delay of half a clock cycle between establishing the address and enabling the memory chips allows for the address bus voltages to settle. These bus lines have a fairly high stray capacitance, and the half-cycle allows for these strays to become charged or discharged. The next delay must allow time for the memory chips to locate the data. There will be varying voltages on the data bus during this time as the memory chips respond, but nothing is gated into the microprocessor until the leading edge of the third clock cycle. The input of data is then shown shortly after this leading edge. Only a very short time is required for the actual reading at this step. The reading action is terminated by the rise in voltage of the MREQ and RD signals, synchronised to the trailing edge of the third clock pulse. The next clock leading edge will then start another cycle.

The write cycle is illustrated in *Figure 6.11*, and uses the same basic three clock pulses. As in the read cycle, the address is put on to the buses just following the leading edge of the first clock pulse. This is the point at which the previous address is changed to the new address, and the usual settling time must be allowed. The MREQ signal then goes low at the trailing edge of the first clock pulse, but the WR (Write) signal does not go low at this time. The reason is that data is to be put on to the data bus from the microprocessor, and the signals on this bus will take time to establish. These signals will appear very shortly after the MREQ signal, but the WR signal does not go low until the trailing edge of the second clock pulse. This enables the memory chips so that the voltages on the data bus, now steady, can be copied into memory. The WR and MREQ signals go high again at the trailing edge of the third clock cycle, so disabling the memory chips again, but the data voltages remain latched on the data bus until the start of the next cycle.

These read/write cycles make use of the programmable registers of the microprocessor, in particular the (main) data accumulator register. Now look at the cycle in which an instruction byte is read into the instruction register, *Figure 6.12*. The address bus is established on the first clock leading edge as before, but another control signal, labelled M1, is also active. This signal goes low each time an instruction word is being read,

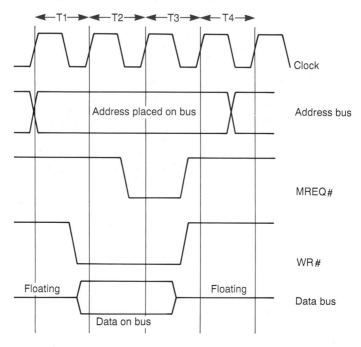

Figure 6.11. The write byte cycle for the Z-80. The data is available on the bus for a longer period than in the read example

and as a consequence the read cycle will need four clock pulses rather than three. The MREQ and RD signals go low at the usual time – the trailing edge of the first clock pulse – but the reading is initiated by the trailing edge of the second clock pulse. The leading edge of the third pulse causes the address on the address bus to change to a 'refresh address', used for refreshing dynamic memory, and the MREQ and RD lines go high again. The instruction word has now been read, and in the next two pulse times it will be analysed so that it can be acted on starting with the fifth clock pulse. Meanwhile, dynamic memory is being refreshed, because the control bus RFSH signal is low. The 'refresh address' uses only the lower 7 bits of the address bus. This is a hangover from the time when $16K \times 1$-bit dynamic memories were standard. The principle is that a 16K memory can be addressed as 128 rows and 128 columns, and each 128 group needs 7 address bits, since $2^7 = 128$. The refresh system is beyond the scope of this chapter, but the principle is that a row access address (RAS) and a column access address (CAS) will be used on the memory chips to read and rewrite data at each addressed cell. Other processors use a separate chip for this purpose. The refreshing operation is carried out

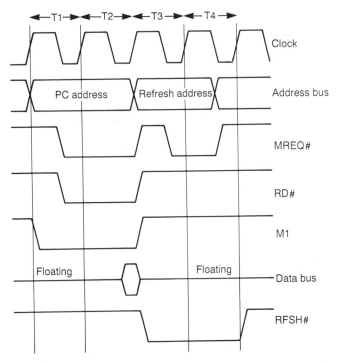

Figure 6.12. Reading an instruction byte into the Z80. The data is placed on the bus near the end of the second clock cycle, and the refresh signals are sent out for the remainder of the cycle

during the time that both the RFSH and MREQ signals are low. For details of the design of memory refresh circuits, you should consult the manuals for dynamic memory chips.

Three-state control

One very important feature of bus lines which has not been brought out so far is 'three-state' control. This does not imply a third logic state voltage, but the ability to float a line. Suppose that a data bus line is at logic 1 and, for some reason, another device connected to the data bus places a 0 signal on the same line. The least that can happen is that the line voltage becomes indeterminate, somewhere between 1 and 0. The more serious problem is that excessive current will flow, and either the microprocessor or the other device will suffer a burn-out, causing a complete system breakdown. This is because a 1 on the data line from the microprocessor implies that the data line is connected to the positive supply voltage through an MOS device. The '0' on the line from the other device means that the line is also connected to earth through another MOS device. With two conducting MOS devices in series across the voltage supply, the probability of burning out one or both is very high.

To avoid such a catastrophe, it must be totally impossible for such a clash of logic voltages to occur in any correctly designed system. Since, by definition, the buses connect the microprocessor to all the other chips that handle logical signals, this is by no means a simple task, but the timing of signals on the buses offers a solution. The solution is to construct the microprocessor, and some other chips (such as buffers) which may be used, with an isolating stage for each pin that connects to a bus. During the time that the microprocessor controls the bus totally, there is connection between the internal microprocessor voltages and the bus pins. At any other time, the isolating stage allows the pins of the buses to float to any voltage that is set by external signals. Because of the isolation, the voltages on the buses at these times do not affect in any way the voltages inside the microprocessor. The signals of the control bus can then determine the times at which external devices can make use of the buses, with the bus pins of the microprocessor floating.

The designer of a microprocessor system must ensure that any other devices which can place voltages on the buses are controlled by the appropriate timing signals so that they will be effective only when they are required. All of the address pins and data pins of a microprocessor are of this three-state type, and several of the control bus lines also. All microprocessors use similar systems of address and data lines, but the signals on the control bus vary considerably from one microprocessor to another, and it is not always obvious which of the control bus signals may be capable of three-state operation.

The control bus

The principles of address bus and data bus are common to all micro-processors, and the only real complication in this respect is the use of multiplexing as on chips such as the Intel 8086. The control bus, however, is something that is very specific to a microprocessor type, and even microprocessors in the same family of devices can differ significantly from each other. At this point, it is important to note the polarity of signals. In the early types of microprocessors, the convention was that a bus line was active low, meaning that the signal took effect when the voltage changed from logic 1 to logic 0. This was invariably indicated in the name of the signal by placing a bar over the abbreviated name, and later by placing the # mark following the name of the signal. More recent designs of microprocessor use a mixture of control signals, some of which are active low, others active high. It is more usual now to find in applications sheets the words 'active' and 'inactive' used for these signals, to avoid having to specify high or low for each one. Great care must therefore be taken to find out which state is the active one, and if you are using a microprocessor type for the first time, it is always a good idea to type out a reminder sheet for yourself.

Some control bus signals are inevitably the same on almost any microprocessor. One obvious example is the use of RD and WR signals to control the reading or writing of RAM. Older types of microprocessor, such as the Motorola 6800 and Mostek 6502, used a single R/W line, which was placed at logic 1 for read and logic 0 for write. Later types tended to use separate pins for RD and WR, so that a state of neither reading nor writing could be more easily arranged. The Z80 uses this system, with the two signals on adjacent pins, both active low, and both three state. This scheme is the normal one, but the modern microprocessor chips tend to use these pins for more than one function.

Another control signal type which is found on all microprocessors, almost regardless of design or manufacturer, is the RESET. The purpose of the RESET signal of the control bus is, as the name suggests, to reset the microprocessor so that the next clock pulse will behave as if it were the first pulse after switch-on. This means that the microprocessor must have in its program counter (also known as instruction pointer) register the first address of a routine which will do whatever needs to be done at switch-on, or at any restart. The design of the microprocessor will determine what address must be used.

The HALT or WAIT action is also one that is implemented on a number of different microprocessors in ways that do not, on the surface, look alike. The more modern microprocessors, in particular, use a HALT action which allows another microprocessor to take over the buses, and this is normally an action that requires a confirming signal. The HALT is started by changing the voltage on one pin, but the buses are not available until the voltage on another pin has changed.

Timing and bus control

The control bus signals will all play some part in timing, and these actions are particularly complicated when the microprocessor shares buses with other microprocessors in the same system. In addition, a complete timing diagram must show the effect of signals such as WAIT, interrupts (see later), and various outputs that are used as acknowledgements. For example, *Figure 6.13* shows the full timing diagram for a Z80 memory read and write sequence. The main actions on the RD and WR control bus lines should already be familiar, but the diagram also shows the WAIT and MREQ signals. The MREQ is an output which is active low. Again, the name is an abbreviation of memory request, and the signal is used to indicate that the address bus holds a valid address for memory, as distinct from a port address (see later). The MREQ signal is synchronised to the

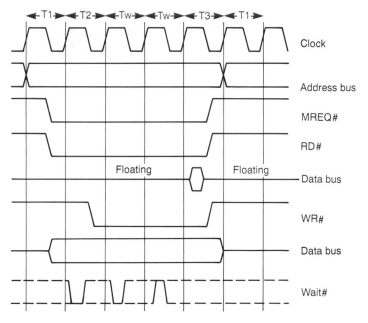

Figure 6.13. The full timing diagram for a Z-80 read and write sequence with Wait signals being used to delay the action. The polarity of the Wait signal determines whether the wait is imposed or not

trailing edge of the first clock pulse in a read or write cycle, and it goes high again at the trailing edge of the third clock pulse. The start of the RD pulse is synchronised with the start of the MREQ pulse, but this is not true of the WR pulse, which starts one clock cycle later than the MREQ pulse. This is necessary because the memory is addressed when the MREQ signal goes low, and data is available on the data bus. If the WR signal was also issued at this point, however, there would be a risk of data being put into false memory addresses during the time that memory was responding to the address. By delaying the start of the WR signal by one clock pulse, the address bus signals can settle, and the memory has time to respond before writing is enabled.

The PC register and addressing

The microprocessor runs a program by outputting address numbers on the address bus so as to select memory. At each memory address, data will be read so as to obtain an instruction or the data to carry out an instruction, or written to store in memory. The sequence of reading memory is normally a simple incrementing order, so that a program which starts at address 0000H will step to 0001H, 0002H, and so on, automatically as each part of the program is executed. The exception is in the case of a jump, caused by an interrupt or by a software instruction. A jump in this sense means that a new address will be placed into the program counter register, and the microprocessor will then read a new instruction starting at this address. For the moment, however, the important point is that the normal action is one of incrementing the memory address each time a program action has been executed.

The program counter (PC) or instruction pointer (IP) register is the main addressing register, connected by gates to the address pins of the microprocessor. The number in this register will be initialised at RESET, and incremented each time an instruction has been executed, or when an instruction calls for another byte. Imagine, for example, that the whole of the RAM memory from address 0000H is filled with an NOP instruction byte. NOP means 'no operation', and its action is simply to do nothing, just go on to the next instruction. If the PC is reset to contain the address 0000H, then the NOP word at this address will be read, decoded, and acted on. The action is actually nil, and so the PC is incremented to address 0001H, the word read, and the action repeated. If the entire memory is filled in this way, the microprocessor will simply cycle through all of the memory addresses until the address reaches 0000H again, and the whole addressing sequence will repeat. The time needed to cycle through memory in this way is very short. For a Z80 using a 4 MHz clock, for example, 64K of memory could be covered in a time of about 65 ms. For the Z80, the NOP instruction word is 00H, so that the addressing of this particular chip can be checked by connecting all data lines to logic 0, and switching on. Other microprocessor types, however, do not necessarily use 00H as the NOP instruction.

Of course, in a real-life system, the memory is not full of NOP bytes. The timing and the PC actions depend very much on what instruction bytes are present, and even more so on the addressing method. Looking at addressing methods brings us into the realm of software, but is necessary for understanding how the PC and buses can be used during an instruction that involves the use of memory. The principle is simple enough – that many of the instructions of the microprocessor require a word (or more) to be obtained from the memory. Carrying out instructions like NOP, or the shift and rotate instruction, do not normally require any load from memory. This is because these actions are carried out on a single byte, word or Dword, which can be stored in one of the registers of the microprocessor. For a lot of actions, though, one word will be stored in a register, and another word must be taken from memory. Since the principles are the same, the illustrations that follow are based on 8-bit processors using single-byte units of data.

Addressing methods

To illustrate addressing methods for an 8-bit processor we can imagine that a byte, 31H, is contained in the main arithmetic (accumulator) register of the microprocessor, and that we want to add to this another byte, 4BH, that will be taken from memory. The addressing method that is used will be determined mainly by software considerations, but the general aims are convenience and speed. The simplest, fastest and most convenient of the addressing methods is immediate addressing. When immediate addressing is used, the word that has to be read from memory is stored immediately following the instruction byte. At the point in a program where we want to carry out the addition of the word 4BH into the accumulator register we place the instruction byte. In the next memory address we place the 4BH byte. The action will be that when the PC increments to the address of this instruction byte, the instruction will be read into the instruction register, and decoded. As a result of decoding, a few clock cycles later, the data word will be fetched from memory simply by incrementing the PC, placing this address on the bus, and carrying out a read action. The bytes 31H and 4BH are then added, and the result 7CH is placed back in the accumulator. Even this simplest addressing method has involved several steps of microprogram and anything from 4 to 16 clock pulses, depending on the microprocessor type. The attractiveness of this method is that the natural incrementing action of the PC is being used to obtain the data. The problem is that the method is not always the most practical. In a system which contained both ROM and RAM, you would normally want to hold a program in the ROM and data in the RAM, but the use of immediate addressing requires the program and data to be mixed in together. This is just one reason why other methods may have to be used.

The main addressing method is called variously direct, extended, or absolute addressing. Suppose that the address of the word 4BH is 7F23H. This address consists of 2 bytes, and can be stored in two consecutive single-byte memory locations. If we store, in sequence, an ADD instruction; and then the 2 bytes of this address, we can make the microprocessor locate the 4BH word from its memory address. This time, the ADD instruction word will be different, because the action that is needed is different. The instruction word is followed in the program memory by the two parts of the address, usually in low–high order. In other words, the address of 7F23H is stored as 23H, and then 7FH. The sequence of actions for the addition is then as follows.

First, the instruction word is read into the instruction register for decoding. At the end of the decoding action, the PC is incremented so as to locate the low byte of the address. When the address bus has settled, this word is read and stored in a special address register. The PC is then incremented again, and with this new address on the bus, the high byte of the address 7F23 is read, and is placed into the address register. The buses are then isolated, and the contents of the address register and the PC are exchanged. This puts the address 7F23H into the PC, and this address is now put on to the address bus. A read action will fetch the word 4BH at this address, and from now on the addition action can take place as before.

This has involved considerable use of the buses, plus an additional register, and an interchange. Finally, before the next instruction can be fetched, the PC and address registers are exchanged again, so that the address in the PC is once again the address of the high byte 7FH. The PC is then incremented to prepare for fetching the next instruction byte. Like the instruction register, the register in which an address is assembled is not available to the software programmer, only to the microprogram.

Extended addressing of this type requires much more bus action, and inevitably takes much longer to execute than the simpler immediate addressing. Most microprocessor designs feature ingenious alternative addressing methods which allow faster addressing for special cases. An action that is very often needed in all kinds of applications is fetching a set of bytes in sequence. A set of bytes stored in order in the memory could be fetched by extended addressing, but it is much more convenient if the address of the first of the bytes can be stored in a register, and the register contents (the address) incremented each time a word is fetched. In this way, the buses are used much less, once to fetch the instruction byte and once to fetch the word from memory, with no need to use two fetch operations to assemble an address each time. Addressing of this type is featured on most modern microprocessors, and was also a feature of the Z80 and its immediate ancestor, the 8080.

In this chapter, details of all the possible addressing methods of each microprocessor would be out of place, since we are dealing primarily with hardware. Nevertheless, the subject is important because it illustrates how extensively the buses may be used for a comparatively simple operation. It is up to the hardware designer to ensure that the buses will operate correctly, with no clashes. The software designer must then devise a program which will fit into the available memory, and carry out its actions as quickly and economically as possible. This will mean organising the data so that the quickest and simplest addressing methods can be used.

Interrupts

An interrupt is a signal that interrupts the normal action of the microprocessor and forces it to do something else, almost always a routine which starts at a different address and which will carry out an action that deals with the needs of the interrupt signal. Such a routine is called an 'interrupt service routine'. This simple description leaves a lot unanswered. For example, suppose that the microprocessor is half-way through an instruction when an interrupt occurs? What then happens to that instruction? How can the microprocessor system resume its normal program actions after an interrupt, when the interrupt service routine has forced the microprocessor to jump to a new address? What happens if another interrupt comes along when the microprocessor is dealing with an interrupt already? We will deal with these points later, and for the moment concentrate on why an interrupt system is used before we examine the details of how it is implemented.

Suppose, for example, that the microprocessor is working in a loop, a repeating set of program instructions carrying out some repetitive action. How would you make certain that pressing a key on the keyboard would cause the microprocessor to find the correct code for that key, and place the result on the screen? One way involves purely software. The loop that the microprocessor is performing must contain a test of the keys which will run a suitable routine if a key happens to be pressed. This is a system called 'polling', and the objection to it is that this test will run each time, even if the keys are pressed only once per minute or so. The result is to make the loop run much less quickly than it would if it contained no key-testing portions. The alternative to this software method is a hardware interrupt. Pressing any of the keys on the keyboard generates an electrical signal which is applied to one of the interrupt pins of the microprocessor. When this signal is received, the microprocessor executes an interrupt routine. In doing so, it will complete the instruction that it is processing, and then jump to an address to get directions for a service routine, in this example the routine that reads the keyboard.

The fact that the current instruction is always completed answers one of the points about interrupts. This interrupt system allows the machine to operate at high speed in its normal processing, without the need to test for a key being pressed until the event happens. A compromise method that is used in several computers is to keep the key test routine separate, and to generate an interrupt 50 times per second (using the field synchronising pulses for the monitor display) to test the keys. The important point in all this is that an interrupt is a signal to a pin, and its effect must be to make a piece of program run. The use of interrupts, then, concerns both hardware and software designers.

The second point, about returning to the correct address, is dealt with automatically. All microprocessors allow a part of the memory to be designated as a 'stack'. This means simply that some addresses are used by the microprocessor for storing register contents, making use of the memory in a very simple last-in–first-out way. Precisely which addresses are used in this way is generally a choice for the software designer. When an interrupt is received, the first part of the action is for the microprocessor to complete the action on which it is engaged. The next item is to store the PC address in the stack memory. This action is forced by the microprogram, so as far as the user is concerned, it is completely automatic. Only the address is stored in this way, however. If the interrupt service routine will change the contents of any other registers of the microprocessor, it will be necessary to save the contents of these registers on the stack also. This is something that has to be attended to by the programmer who writes the interrupt service routine, saving the register contents at the start of the interrupt service routine, and replacing them afterwards. Finally, the problem of multiple interrupts is dealt with by disabling the interrupt mechanism while an interrupt is being serviced. This is not necessarily automatic, and will often form one of the first items in the interrupt service routine. The tendency over the short history of microprocessors has been to delegate these actions to the software writer rather than to embed them in hardware.

Most microprocessors can make use of more than one type of interrupt signal. The two main types are maskable and non-maskable. A maskable interrupt is one that can be enabled or disabled by software instructions. This allows the software designer to disable interrupts at times when an interrupt would cause corruption of data. For example, when data is being loaded from a disk or stored on to a disk, the transfer of data is a strictly timed operation. If pressing a key could interrupt this, the memory of the machine could contain a fraction of a program and the disk controller could be left midway through a sequence of actions. A software designer would therefore want to disable any interrupts while a disk load (or save) was being executed. There may, however, be events (such as pressing a special BREAK key) which must cause an interrupt, and this need is catered for by a non-maskable interrupt. The normal method of dealing with this double system is to have two separate interrupt pins on the body of the microprocessor.

Inputs and outputs

A system which consisted of nothing more than a microprocessor and memory could not do much more than mumble to itself. Every useful system must have some method of passing bytes out of the system to external devices, and also into the system from external devices. For computer use, this means at the very least the use of a VDU screen and a keyboard, however primitive. For modern systems, you can add to this a disk drive, various sockets where other add-ons can be plugged (such as light pens, joysticks, a mouse, trackerball, etc.), and possibly inputs for measuring voltages. If the microprocessor system is intended for machine-control purposes, then the outputs will be very important indeed, because these will be the signals that will control the machine. The inputs may be instruction codes from a disk or cassette, pulses from limit switches, digital voltage readings from measuring instruments, and so on. Whatever the function of the system, then, the inputs and outputs form a very important part of it all. A surprising number of control actions of a microprocessor system, in fact, consists of little more than passing an input signal to an output device, perhaps with some monitoring or comparison action thrown in.

Before we get involved with the details of how signals are passed between a microprocessor system and the rest of the world, we need to be aware of the problems that are involved. The main problem is one of timing. The microprocessor system works fast, governed by the rate of the system clock. There is no point in having, in the software, instructions that will make the microprocessor send a word out of the system unless you can be sure that whatever you are sending the word to can deal with it at that time. The time that is involved might be perhaps a couple of clock cycles, a fraction of a microsecond, and there are not many systems, apart from another microprocessor, that can deal with such short-duration signals. The same problem applies to incoming signals. At the instant when you press a key on a keyboard, can you be certain that a microprocessor is executing an instruction that will read the key? If the instructions for carrying out a read of the keyboard are carried out in a loop, then the situation is easier, but the software will have to see to it that the key is not read each time the loop is run, because otherwise you might simply find the system reading the same key several million times until you released it. The timing of these input and output signals is therefore a matter of priority. Another matter is the nature of the signals. The signals from a microprocessor system will be digital signals, using standard TTL voltages. Some of these signals may be active low, others active high. If the input from another system happens to be an analogue signal of maximum amplitude 50 mV, or if the signal that is required by a device at the output is a ± 50 V 500 Hz AC signal, then a lot of thought will have to be devoted to interfacing. This is outside the scope of this chapter, because for machine-control applications in particular, interfacing is by far the most difficult action to achieve in a new design. Some forms of interfacing are dealt with in Chapter 7.

Ports

A port is a circuit which controls the transfer of signals into or out of the microprocessor system. If the only requirement for a system is to pass a single bit in and out, the obvious method is to use a simple CMOS or LSTTL latch chip, connecting with one of the data lines of the microprocessor system. Even if eight data lines are needed, a straightforward hardware latch may be sufficient, particularly if the signals are all in one direction, or if most of the signals are in one direction. The port chips which are available for use with specific microprocessors are designed to provide all solutions to all users, and are therefore very complicated programmable devices, requiring software as well as hardware. If the system is complicated enough to warrant the use of a port chip, or if the requirements for inputs and outputs are likely to grow, then the use of a port chip is justified. For a lot of systems, however, a port chip would be an unnecessary indulgence, carrying out the work of a few simple latches at the cost of a lot of software and hardware complications. We will look first, therefore, at how simple latch systems can be implemented.

Whether data has to be read in or written out, any latch or port must have an address. In other words, a unique set of bits on the address lines must be able to release the latch to allow data bits to be stored. For a small system, the ordinary memory-addressing system can be used for this. Suppose, for example, that a small microprocessor system uses ROM and RAM in a total of 16K of memory, and that the microprocessor is a standard 8-bit type, with 16 address lines. To address 16K requires only 14 of these lines, leaving A14 and A15 free. Now these two pins can be used to enable latches. You could, for example, use A15 to enable the reading latch and A14 to enable the writing latch. Any address in the range 4000H to 7FFFH would then enable the writing latch, and any address in the range 8000H to BFFFH would enable the reading latch. Addresses in the range C000H to FFFFH would enable both latches. If you are uncertain how these hex numbers are obtained, the binary equivalents are shown in *Figure 6.14*. The data lines would be connected to the inputs of the writing latches and the outputs of the reading latches. This requires the use of latches which can be put into a third floating state, as most latches can. Writing data then requires the software for a memory write, addressed in the correct range, and reading requires a memory read instruction, again using the correct address range. The fact that any of a very large range of addresses can be used is not particularly important in a small system, unless there is a possibility of expanding the system at a later date.

For a larger system, in which all of the memory lines may be in use, other methods are available. One is complete address decoding. The latch is mapped to an address, perhaps FFFFH. Mapping implies that this address, and only this address, will allow the latch to be enabled, so that this particular state (a 1 on each address line) must be detected. This can be done by using AND gates, or more practically, with 8-input NAND

Hex address	Binary address
4000H	0100000000000000
7FFFH	0111111111111111 (A15 low, A14 high)
8000H	1000000000000000
BFFFH	1011111111111111 (A15 high, A14 low)
C000H	1100000000000000
FFFFH	1111111111111111 (A15 high, A14 high)

Figure 6.14. Binary and hex number equivalents for some important addresses in the examples

gates. An alternative is the use of bus comparators. These are chips which act as multiple gates, with lines of a bus connected to one set of pins and fixed voltages on another set. The output of the comparator is a logic 1 only when the corresponding pins are at identical voltages. One set of decoders will be needed for each address that has to be memory mapped in this way, but careful planning may allow one main decoding section, with lower-order address lines forming the final gate inputs, as *Figure 6.15* illustrates.

The use of complete address decoding implies, on an 8-bit micro-processor, the decoding of all 16 address lines. Though this allows a wide choice of addresses, and a large number of possible latches, it does generally mean that all 16 address lines have to be decoded. The alternative is 'port' decoding, which is provided in all microprocessors. Port decoding uses only a portion of the address bus, for example the lower eight lines of a 16-line address bus. This would permit 256 latches to be used, more than will ever be needed by most systems. Obviously, since these lower eight lines of an address bus are continually being used in addressing memory, some method is needed to distinguish the use for latching. This is done by means of the microprocessor's IN and OUT instructions. These software instructions cause a control bus signal to be used to distinguish a port operation from a memory read or write, so that the latching system must be enabled by the combination of this port

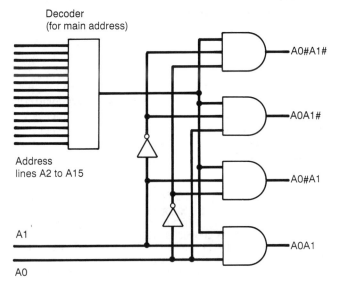

Figure 6.15. A decoding system for ports, using a full address plus two low-order address lines

signal and the decoded address lines. The Z80, for example, uses its IOREQ line in this way. The IOREQ line is taken low when an IN or OUT instruction is executed, and this line signal, in conjunction with the bits on the lower eight lines of the address bus, enables a litach or port. This IOREQ line also goes low when an interrupt occurs, allowing a data word to be read from a port. With only eight address lines to decode, plus a control signal, the construction of a latch circuit becomes considerably easier.

Port chips have been developed to provide a more universal solution to all of the interfacing problems that are likely to arise with microprocessor systems. As such, some of the port chips that are available are very complicated devices, as complicated as the microprocessor chip itself, and nearly always programmable. The use of a programmable port implies that its use must be a matter that is jointly planned by both the software and the hardware engineer. Simple solutions are often more efficient, and it is better to knock up a working system from gates in an hour than to spend 3 weeks trying to understand how a port chip can be used. The nature of microprocessor systems often tempts a designer to make use of a compatible port with a microprocessor chip, but this may not necessarily be the most cost-effective answer to the problems of a small system. In this section, then, we shall deal with ports starting with the simplest chips that are available.

One of the simplest general-purpose ports is the Signetics 8T31. This consists of two sets of data input/output pins, each of which is described as a port in the applications sheets. One set, labelled IV0 to IV7, connects the chip to the data bus of an 8-bit microprocessor. The other set, labelled UD0 to UD7, is the 'user port', the connections to external devices. The data can be passed in either direction, and latched, with signal level inversion if the data is passed from an IV to a UD pin, or in the opposite direction. The chip is in a 24-pin package.

Figure 6.16 illustrates the pin allocation of this chip, with its controlling signals. The master enable (ME) pin allows the port to be gated by any form of address selection method that is used. It controls reading and writing of the latched data from the microprocessor system, and does not prevent input and output at the user side of the port, only at the microprocessor side. In this connection, the port has a useful initialisation feature. Providing that the clock input is maintained low until the supply voltage has reached at least 3.5 V, the ports will be initialised with all of the user-port pins at logic 1 and all of the microprocessor pins at logic 0. With the port operational the reading and writing of data into the user side is controlled by the BIC and BOC pins, each of which is active low. The BIC pin is the input control which enables data to be written into the latch from the user lines. The BOC pin voltage, also active low, permits the latched data to be read from the user side. With both of these pins high, the user data pins are floated. Data can be output at any

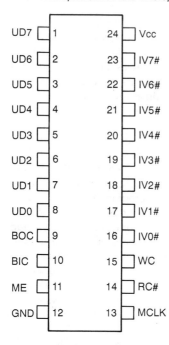

UD7	1	24	Vcc
UD6	2	23	IV7#
UD5	3	22	IV6#
UD4	4	21	IV5#
UD3	5	20	IV4#
UD2	6	19	IV3#
UD1	7	18	IV2#
UD0	8	17	IV1#
BOC	9	16	IV0#
BIC	10	15	WC
ME	11	14	RC#
GND	12	13	MCLK

Figure 6.16. The pin allocation for the Signetics 8T31 port chip, a simple general-purpose port

clock voltage, but the input of data from the user side is at the clock high level only.

The microprocessor side is controlled by the master enable, as noted above, and also by the WC and RC pins, strobed by the clock input. When the WC pin voltage is high, data is written into the latch from the microprocessor system at the clock high level. When the RC pin is low, data can be read from the latch into the microprocessor system at either clock level. Conflict in the data latch operation is prevented by giving the user side priority. In other words, if BIC is enabled to read from the user side into the latch, no reading can be accepted from the microprocessor side at the same time. Apart from this, the two sides operate independently. There is no direct provision for an interrupt signal to the microprocessor from this port, however, and if an interrupt is needed, it will have to be provided by a separate flip-flop.

The 8T31 and others in its family provide very useful and simple port facilities in conjunction with any method of port mapping. The 8T31, incidentally, allows internal programming of a decode address so that a specific number on the data lines, plus an enabling signal, will enable the chip for following data. For machine-control applications in particular, the use of such ports may be a very straightforward answer to porting problems. Remember that this port is an 8-bit type, and very often machine-control applications may not require many bits. There is no reason for all the bits of a port to form part of a single byte, because if necessary each bit can act as an independent switch, or a group of bits can be interpreted and used as switching control bits. The interpretation of how the bits are arranged then becomes a problem for the software engineer. For the larger systems, though, a full-scale port chip will have to be used, and these devices demand a considerable degree of cooperation between software and hardware engineers, both in planning and in testing.

Keyboard interfacing

One form of interface which is needed in practically every microprocessor system is a keyboard port. The form that this takes will depend on what size of keyboard is to be used. For a number of machine-control systems, a simple hexadecimal keyboard will be all that is needed for commands which can be controlled by number or number sequence. For computer use, of course, a full keyboard of around 60 keys will be needed, and in practice the use of 'function' keys and separate number keypads can push this number above 70. All of these keys will have to be interfaced with the system in such a way that each key will give rise to a distinct and separate code. The most practical method of achieving this is the use of a key matrix, the principle of which is illustrated in *Figure 6.17* for a set of 16

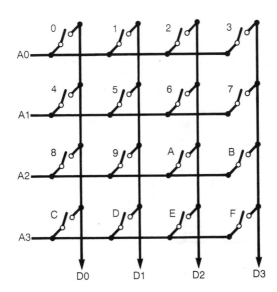

Figure 6.17. A simple key-matrix system, in this case for a set of 16 keys only, using row and column signals

keys, illustrated as a hex set (with no RETURN key shown). Each key makes a connection between two lines, shown in the diagram as A0–A3 and D0–D3, so that in this example the 16 keys are serviced with eight lines. By using eight lines in each group, 64 keys could be serviced, and a 16 × 8 matrix, using 16 address lines and eight data lines, could operate a matrix of 128 keys. The matrix principle therefore allows for as large a keyboard as is likely ever to be used. The use of numbers such as 16 and 8 does not imply that this number of lines need be taken from the system, because the decoding of four binary lines will give 16 outputs, and eight signals can be coded as three binary lines. For a simple system, however, the use of address lines for the A0–A3 lines of *Figure 6.17* and data lines to a port for the D0–D3 outputs would be acceptable. The software of the system must test the port, either on the basis of polling or by the generation of an interrupt. If an interrupt is to be generated when any key is pressed, then each of the D lines must be taken to an OR gate (*Figure 6.18*) in order to detect an output on any line. The principle of programming is that successive lines in the A set are activated, and the D lines tested for any return. The unique combination of digits on the two sets of lines will constitute the internal code number for one key, and this can be converted by the software into whatever is needed within the system.

The more elaborate systems can use a single port, with a 64-key matrix (arranged as 8 × 8) serviced by just six lines. The lines D0–D2, for example, can be decoded to provide bits on eight lines, and the output eight lines taken to an encoder to give just three lines of return signals on D3–D5. If the port is connected so that the output lines from the system, labelled as D0–D2, are supplied from address lines which will be cycling continually, then pressing any key will give a signal on one of the output lines, and this can be gated and latched to provide an interrupt. The service routine can then read the port word as the code for the key that has been pressed. Most keyboards allow for at least one key which is not wired into this matrix arrangement, and which will perform an interrupt of a higher priority than the other keys.

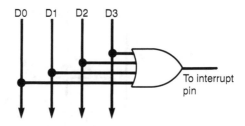

Figure 6.18. Using an OR-gate circuit to detect if any keys are pressed so as to generate an interrupt

Video interfacing

Outputs from a very simple machine-control system to displays of the seven-segment type are straightforward, and can make use of the various seven-segment decoders and drivers that have been available for some considerable time. Display on a video monitor would be considerably more difficult to design from scratch, but fortunately this is never necessary unless the requirements are very unusual. Providing that standard television practice is followed by the monitor, chips exist to make the interfacing relatively simple in hardware terms at least. It helps, however, if you know just what these chips are doing, particularly if for any reason any one of them is not working according to plan.

As an example, consider what is required to display the letter 'A' at the centre of one character line on the screen, the second character line down. We will suppose that the screen can display 80 characters in each line, and 25 lines of characters from top to bottom, since this is a very common requirement. We will assume also that the monitor uses a 625-line type of TV signal, with no interlacing. This is no place to start explaining TV techniques, and if you have not come across terms like scanning and interlacing previously, I must respectfully refer you to a book on TV techniques. Given these essentials, then, the display of the letter 'A' requires a set of actions, some of which will be carried out by the main microprocessor system, and others by the video interface system. The dividing line will have to be positioned by the designers of the system, but the example I shall use is one that is followed by a large number of designers.

The letter 'A' is coded in ASCII as the number 65 denary, 41 hexadecimal. To display the shape of the letter, however, a set of codes will be needed, and the normal method of producing these codes involves the use of a character matrix. For example, each character may be represented (rather poorly) by filling in a matrix of 8×8 squares (*Figure 6.19*). Since characters on the screen must appear separated from each other, only seven of the matrix lines and seven columns of such an 8×8 matrix will be used for a character, and *Figure 6.20* shows how the letter 'A' would be represented. A matrix of this size is not really sufficient for a full range of characters, because the lower-case letters need to be displayed on a smaller part of the matrix, with space for the descending tails of letters such as g, y, p and q. For this reason, larger matrices, such as 15×12, are normally used for character blocks, particularly for the better-quality display that is expected of a 'business' computer that will always be used with 80-character or 132-character lines. For this example, however, we will stay with the principle of the 8×8 matrix. If we are to have some 600 scan lines on screen and 25 lines of characters, then the depth of each character block must correspond to 24 scan lines, three scan lines per matrix line. This gives characters that are rather too

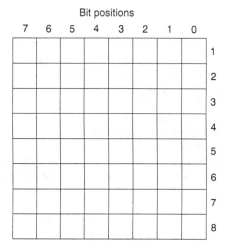

Figure 6.19. An 8 × 8 matrix for character generation

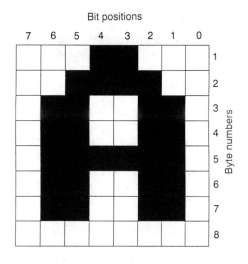

Figure 6.20. The letter A as it would appear in 8 × 8 matrix format

elongated in comparison with their width, and adjustments may be needed. For 40 character per line work, the 8×8 matrix is quite well suited, however.

The character matrix gives rise to a set of 8 bytes, one per line, to describe the character shape. This conversion is normally undertaken by the main microprocessor system, using a 'character-generator ROM'. Early computers kept the character shape codes exclusively in the ROM, but on modern machines it is more usual to load from ROM into RAM at the time that the machine is powered up. In this way, a selection of letters for different languages can be kept in the ROM, and one set chosen for use. More important, with the character block shape numbers in RAM, the user can change character shapes, or design additional characters for special purposes. In either case, the software of the main microprocessor system will convert from the ASCII code of the character into the character block number. In this example, one word of each one will always be zero to ensure separation of lines of characters.

The next part of the action requires the use of memory and some processing action. On the simplest systems, the main microprocessor system may be used to provide the display on the screen, but if this is done, then the screen will go blank while the microprocessor is otherwise engaged. On modern systems, it is more common to make use of a separate video controller chip to carry out the processing actions. The main system will have supplied the set of bytes that represents the character shape, and also two numbers which locate the word on the screen. These will be a column number (0 to 79 for the 80-column screen display) and the line number, which will be 0–24 for the conventional 25-line screen. If each of these is provided as a byte, then 10 bytes in all are being used to specify each character on the screen. This does not mean that 10 bytes must be stored for each character. If the screen is 'memory mapped', meaning that the position of each part of the character on the screen is determined by its address in a block of memory, then the row and column position numbers are used to place the character block bytes into the correct addresses in this screen memory, and then play no further part in the display action. Once again, this is the normal system for computers, though the relationship between the memory address of a word and the position of the corresponding dot on the screen may range from the simple to the very complex. The simplest method is a direct geometrical mapping, in which each consecutive portion of the screen corresponds to one of a set of consecutive addresses in memory, and each bit in a word corresponds to a dot (a pixel) of a character matrix on the screen. For example, with an 80-character line and an 8×8 matrix, there would be 80 bytes per line and 25 lines, giving 2000 memory locations. This is a very modest requirement by modern standards, and many computers allocate 16K or more for screen display, particularly if colour or high-resolution graphics are to be used.

With the character block bytes in the screen memory, the main microprocessor system is normally relieved of further responsibility for display, until the screen has to be updated or scrolled. Disregarding these complications, we want to see now what the video chip has to do. First of all, the video chip has to generate the synchronising signals that will keep a TV type of display locked to the signals from the computer. Not all video chips do this equally well, and a few home computers have been notorious for their inability to produce a locked picture on certain makes of TV receiver, particularly when corners have been cut in the design of the video sections. When a monitor is used, the video chip should be one that will produce good clean synchronisation for the type of signal that the monitor must use. The details of this depend on whether the monitor will be monochrome or colour, and if a colour monitor is used, whether the signals will be separate RGB, or composite. The normal scheme is to use composite signals (video plus synchronising pulses) for monochrome monitors or for TV modulation, and separate RGB plus sync. for good-quality colour monitors. For the sake of this example, we shall assume a monochrome monitor and composite video signals. The video chip must then act as a bit reader and timer, generating video pulses from the bytes of data in the screen memory.

Consider, for example, the letter 'A' at the centre of the second character line. If this is all that is to be displayed on the screen, then the screen memory will contain eight lines of zero bytes, positions 0 to 79 on each of the first eight 'lines' of the character shape top line. Since each line of character block corresponds to three scanned lines, the video interface chip will read a set of 80 memory locations three times at a speed which will cover the width of the screen in the correct time, about 64 μs for a 625-line display. Since all the bytes are zero, no signals will be sent out. On the next eight sets of memory locations, however, the centre character is 'A'. This implies that one dot will be sent out to the middle of the first three screen lines of this group. As the video chip reads the first set of 80 bytes, then, it will come to one which contains a single '1' bit. When the video interface chip reads this, it generates a pulse of about 100 ns. The arithmetic of this is straightforward. If a complete line of 80 characters is being covered in 64 μs, then each character takes up 64/80 μs, which is 0.8 μs, and each dot in a matrix that is eight dots wide needs 0.1 μs. Note that this requires fast memory reading, and a wide bandwidth in the video circuits of the monitor for 80-character display. The normal video bandwidth for a monochrome monitor is around 18 MHz, and the use of a TV receiver for 80-column displays is quite inadequate because of the usual 5.5 MHz video bandwidth of a TV receiver. Even the combined TV/monitor sets seldom have bandwidths that are adequate for 80-column work, and very few colour monitors are really well suited. The position of the first (top) dot of the letter 'A' on the screen is determined by the rate at which the video chip reads along the memory addresses, because the address at which the dot is stored has determined its position in the character line. This set of memory addresses will be scanned three times, and then the video chip will turn its attention to the next set of addresses, which contain the data for the next part of the character. The whole of the video memory will be read in this way at the usual frame rate of 25 per second so that each part of the screen can be maintained.

From this brief, simplified description of the simplest mapping arrangement, you will have gathered that the video interface chip will need a clock signal, which will be geared to the requirements of the monitor rather than to the computer. One solution that is often adopted is to use a master clock oscillator whose frequency is suitable for the video chip rather than for the computer. The computer clock is then obtained by dividing down this master clock frequency. Modern video interface chips are, like parallel ports, microprocessor systems in miniature, often provided with registers that allow a great deal of software control over the character size and block pattern. A full description of such a chip would require a book to itself. Along with the complications of the video chip, the memory that is used for video display may be organised in a way quite different to the simple scheme described above. For example, a set of consecutive memory addresses may represent the vertical scanning of a character line, rather than the horizontal scanning that we assumed above. In addition, the bits in each word may not be bits of the same character. A word may represent four pixels, with 2 bits per pixel, allowing four colour codings for each pixel. These complications normally affect the software designer much more than the designer of hardware, but unless a standard type of video chip is used, the hardware designer will be heavily involved at a very early stage. This is because the alternative to a standard chip set is a combination of standard chips and ULAs, and the hardware designer will be responsible for the ULA. For the service engineer, the ULA is a problem because information is often difficult to obtain. Information sheets on standard chips can be obtained easily from the manufacturers, but when a ULA has been used in a circuit, the details of its hardware and software operation are obtainable only from the computer manufacturer. Some manufacturers are very reluctant to disclose any information about ULA operation, and this can be a considerable hindrance to servicing work. If you have any choice in the matter, service only computers for which adequate information is freely available!

Further reading

For details of microprocessor types, see the following Newnes Pocket-books and other Butterworth-Heinemann titles:

8086 Family (8086 to 80486) Pocket Book
Z80 Pocket Book
68000 Pocket Book
Microprocessor Pocket Book

Other hardware

Data Communications Pocket Book
Hard Disk Pocket Book
PC Users Pocket Book
PC Users Companion
Mac Users Pocket Book
PC Printers Pocket Book
Computer Engineers Pocket Book
The Scanner Handbook

Programming Software

MS-DOS Pocket Book
C Pocket Book
Unix Pocket Book

General

Windows-3 Pocket Book
Computer Science (Ian Sinclair)

Assembly language for 8086

Starting MS-DOS Assembler (Ian Sinclair) published by Sigma Press,
Wilmslow, Cheshire.

Chapter 7

Transferring Digital Data

Serial and parallel

Some applications of digital circuitry make use of the digital data at the time when it is obtained. A digital voltmeter, for example, displays the obtained data as soon as the data is collected, and refreshes this data at intervals of, typically, half a second. Even in this case, however, there is some storage in the sense that the display retains the reading until a new reading is made. The storage is transitory, however, using latches, and there is no requirement to transfer digital data from one piece of equipment to another. In many other applications, however, and particularly for computing, data has to be transferred over distances that range from a metre or less up to the maximum distance that a radio signal can reach. In this chapter, we shall look at data transfer methods.

The simplest method of transferring digital data is to connect to a microprocessor bus, usually by way of buffer or driver circuits. The word 'buffer' is used here in the electronics sense of a circuit that acts as an impedance transformer, reducing loading on the source. The computing term means a piece of memory used to gather up data until it is needed. The transfer of data from one board to another in a digital system makes use of the microprocessor bus either directly or by way of buffer circuits, the *bus drivers*. In such connections, all of the microprocessor signals are transferred, including data lines, address lines and all of the synchronising and timing lines.

The more difficult requirement is the transfer of digital data between different pieces of equipment, of which the most common example in computing is the use of a printer. For industrial and instrumentation purposes, the requirements are much more varied but the basic methods are much the same. The choice is of either parallel or serial transmission and reception, and in many cases only serial transmission is possible. No matter which method is used, some form of synchronisation will be needed because the rate at which data can be received by a device such as a printer is never as fast as the rate at which it can be transmitted from a microprocessor. Both serial and parallel data transfer systems must therefore ensure synchronisation of transmission and reception, and the problem is more acute for serial links. The signals that are used for this purpose are called *handshaking* signals.

Parallel transmission means that all the data lines of the microprocessor bus, or a set of data lines, will be used, along with a few synchronising lines. For instrumentation purposes, a more complete set of signals will be needed than is the case for a computer printer. Many of the microprocessors used in industrial equipment are of the 8-bit variety, and for parallel transmission of data all eight data lines will be used. In applications which use 16-bit microprocessors, industrial requirements will usually call for all 16 data lines to be used in a parallel connection, but computer printers require only eight lines at most, and some use only seven.

The difference is owing to the way that data is used. Transfer of data for instrumentation purposes will generally require as many data lines as the microprocessor itself can use, because the nature of the data will require the use of all the lines. Printable characters in computing applications make use of the ASCII code (see *Figure 5.22*) whose numbers can be expressed by a 7-bit binary code, so that only seven lines of data are ever required for a connection to a printer for text. Some printers make use of (non-standard) codes in the range 128 to 255, and these require 8 binary bits and, hence, eight data lines.

The main problems of parallel data transfer are of line length and pulse frequency. These two problems are interconnected because they both arise from the stray capacitances between the leads of the cable. A parallel cable will be driven from a low impedance source (the usual double emitter–follower type of circuit) and will connect in to a comparatively high impedance at the receiver end. The stray capacitance between leads, together with the very fast rise and fall times of the pulses, can therefore induce a 1 signal in a line which should be at level 0. The longer the line, the greater the induced signal, until the voltage becomes great enough to drive the receiver circuit, at which point a false signal will be received.

The practical effect is to restrict parallel printer leads from computers to 1–2 metres. Greater lengths can be obtained by using correctly matched 50 Ω lines, but these are rare in computing applications though fairly common for instrumentation applications. Long parallel links can be used if repeaters are connected at intervals. The repeater consists of Schmitt trigger circuits which will give a 1 output only for signals which reach a preset minimum level, and a 0 output for all others, combined with a very low output impedance. Repeaters for computer printer leads are comparatively simple (but only recently available), because most of the signals are in one direction only – from computer to printer – but for instrumentation purposes the repeaters will normally be required to cope with signals in either direction.

Table 7.1 shows the signals that are used in a Centronics printer output, in this case from a computer that is IBM compatible. Not all printers make use of all of these signals, nor do all computers, but the Centronics standard is sufficiently flexible to ensure that any computer that provides a parallel printer output to Centronics standard can be matched to any printer with a Centronics input. There are various minor deviations between printers and computers, all of which can be dealt with by omitting one or more links in the cable. Plugs and sockets are standardised only at the printer end of the cable which uses a 36-pin plug of the Amphenol 57-30360 type. The plug and socket that is used at the computer end of the cable is not standardised, and small computers in particular seem to use any plug/socket that comes to hand. The IBM-compatible machines use a 25-pin D-connector for both the parallel output and the serial input/output, with a socket used for the serial connection and a plug for the parallel printer connection.

Table 7.1. THE SIGNALS AT THE CENTRONICS OUTPUT OF A TYPICAL COMPUTER

Signal pin	Return pin	Name	Notes
1	19	STROBE	Low to send data
2	20	D0	
3	21	D1	
4	22	D2	
5	23	D3	
6	24	D4	Data bus
7	25	D5	
8	26	D6	
9	27	D7	
10	28	ACKNOWLEDGE	Low when printer ready
11	29	BUSY	High when printer not ready
12	30	PE	High when out of paper
13	—	SELECT	On/off line (out)
14	—	AUTOFEED	
15	32	ERROR	Out of paper, off line, error
16	—	INIT	Reset
17	—	SLCT IN	On/off line(in)
18	—	NC	

Only about 22 pins of the 36-pin connector are likely to be used, and on the 36-pin connector, pins 2 to 9 inclusive are used for the eight data lines of the data bus, D0 to D7. The Centronics standard provides for signal return lines, each at signal ground level, so that twisted pairs of signal/ground return wires can be used. Return pins 20 to 27 inclusive are used in this way for data, but this provision is not always used, particularly for short cables. Pin 1 (return pin 19) handles a strobe signal which is driven low by the computer in order to send data to the printer. The width of the strobe pulse must be at least 0.5 μs, and this pulse is one of three that forms the main handshaking provisions in this type of interface. The other two are BUSY (pin 11, return on 29) and ACKNOWLEDGE (pin 10, return on 28).

The BUSY signal is a steady-level signal that is set high by the printer to indicate that no more data can be received. The BUSY line is taken high when data starts to be entered, during printing and when the printer is off-line, or disabled because of a fault. Most printers contain buffer memory which can range from one line to several pages of printed characters, and the BUSY signal is taken high when this buffer is full. For such a printer, transmission of signals is intermittent because of the time needed to fill the buffer at the normal rate of parallel transmission, which is as fast as the signals can be clocked (subject to the minimum pulse widths that can be used). The use of a printer buffer allows the computer to be used during a print out, and some computers provide buffering in their own memory to assist this detachment of printing from other operations. The snag is that if you want to stop the printer you cannot do so immediately by stopping data being sent out from the computer because the printer will stop only when the buffer is empty. You can, of course, switch off or reset the printer, but this will empty the buffer and you will need to retransmit this data when the printer is ready for use again. The better option is to use the off line switch.

The ACKNOWLEDGE pulse signal is sent out by the printer to indicate that the printer has received data and is now ready to receive more data; usually when the buffer is empty. The relationship of these signals to each other is fairly flexible, as the timing diagrams of *Figure 7.1* indicate. These have been taken from a computer manual and two printer manuals, and they show noticeable differences in the way that the pulses are related, though all are within the tolerance of the Centronics standards. The important point is the maintenance of the 0.5 μs minimum pulse width and timing intervals, so that the clock rate of output is usually considerably lower than the clock rate for the computer itself.

The data, strobe, BUSY and ACKNOWLEDGE signals are the most important parts of the Centronics interfacing, and the remaining signals and their uses are summarised in *Table 7.2*. There is some minor variation between printers in the use of these signals. Not all printers, for example, make use of the Autofeed signal, but such variations are not generally important unless you are trying to use one printer with a cable that was intended for another. It is very unusual to find major problems of compatibility between computers and printers using the Centronics

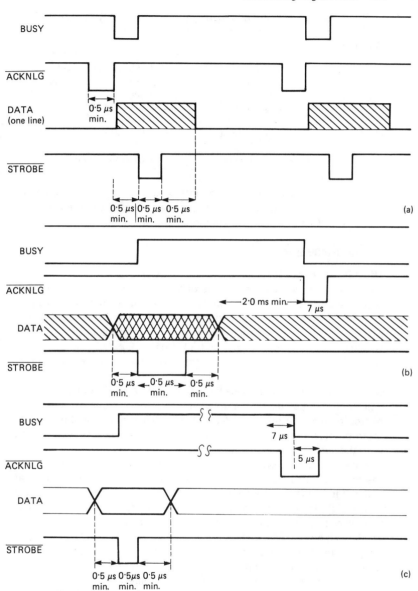

Figure 7.1. Typical timing specifications. (a) Computer, (b) Daisywheel printer, (c) Dot-matrix printer

Table 7.2.   THE OTHER SIGNALS OF THE CENTRONICS SET

Signal	Direction	Use
PE	From printer	High when out of paper
AUTOFEEDXT	To printer	Low to feed one line after printing
INIT	To printer	Low to reset printer and clear buffer. Width $>50\,\mu s$
ERROR	From printer	Low to indicate no paper, off line error
SLCT IN	To printer	Low to allow data to be sent

interface – one example in the past was owing to the computer manufacturer earthing pin 14 so that printers which used the Autofeed signal were forced to take an additional line spacing.

The IEEE-488 bus

The IEEE-488 bus is a parallel data transfer system for connecting complete systems rather than parts of systems and it is widely used in digital electronic instrumentation. The standard dates back to 1974 and much of the detail of the standard is owing to Hewlett-Packard, who hold patents on the handshaking method. Because of this, a licence must be purchased if this handshaking method is used.

The bus is used to connect devices that can carry out actions described as controlling, listening and talking. A device might carry out just one of the functions, any two of these functions, or all three. A controller device will control other devices and is almost always a microcomputer or microprocessor-based controller system. A talker device will place data on to the bus, but does not receive data, and a listener will receive data from the bus but does not place any data on the bus. A counter might, for example, be connected as a talker, placing the data from its count on to the bus but not receiving any data (though it would obey command signals) from the bus. By contrast, a signal generator might be used as a listener, generating signals as commanded by data read from the bus, though not placing any digital signals on to the bus. Many devices will be used as talkers and listeners, receiving signals from the bus (for changing range or function) and placing signals on to the bus to indicate readings. Since the IEEE-488 is primarily intended for instrumentation, the prime example of a talker/listener is a digital multimeter.

The bus, like the Centronics parallel system, consists mainly of data lines with no address information and uses a total of 16 lines on a 24-pin connector. Of these, eight are data lines that are bidirectional, five are bus control lines, and three are handshaking lines. *Figure 7.2* shows the standard pin layout, with data on lines 1 to 4 and 13 to 16 inclusive. The handshaking lines are on pins 6, 7, and 8, with the 'transfer-control' lines

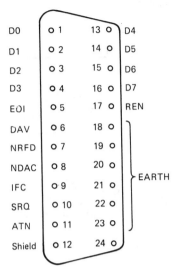

Figure 7.2.   The IEEE-488 connector pinout

on 5, 9 to 11, and 17. The handshaking lines use open-collector outputs, active low, so that these lines can be connected to other output lines (wired-OR connection) without risk to the internal circuitry.

The handshaking lines are DAV (data valid) on pin 6, NRFD (not ready for data) on pin 7 and NDAC (no data accepted) on pin 8. The DAV signal is sent out by a talker device to signal that data has been placed on the data lines and is valid for use. The other two lines are controlled by listeners, with NRFD signifying not ready for accepting the data, and NDAC signifying that data has not been read. When both NRFD and NDAC lines go high, the data is read. The action for a single talker and listener is as shown in *Figure 7.3.* The DAV line from the talker remains high even in the presence of data until the NRFD signal goes high. This is not such a simple action when several listeners are present because the NRFD line is ANDed; it cannot go high until all listening devices are ready. When the NRFD line from the listener(s) goes high, the talker activates the DAV line (low state) so that data can be transferred. The data transfer is complete when the NDAC line rises to the high level and the rate of transfer is controlled by the slowest listener. Typical maximum rates range between 50 and 250 kilobytes per second if the listeners are fast-acting devices.

The bus control lines are used to determine how devices interact with the controller. The simplest of these is the interface clear on pin 9, taken low to reset the system in preparation for use. By contrast, the end or identify (EOI) signal on pin 5 is used to indicate that data transfer is

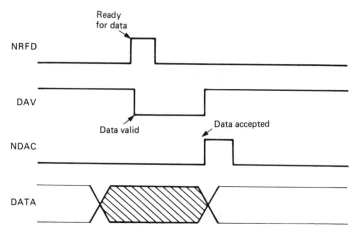

Figure 7.3. The timing of a single talker–listener IEEE-488 exchange. Where more than one listener exists, data is not valid until all listeners are ready

complete. The attention (ATN) line, pin 11, decides the use of the eight data lines. When this pin voltage is low, all eight data lines are used for data, which need not necessarily use all eight data lines. When the ATN pin voltage is high, the lower lines of the data bus are used to hold an address number to which a specific device will respond. In the absence of such a specific address, all listening devices can receive signals sent over the data lines.

The other two bus control signals are service request (SRQ) and remote enable (REN). The SRQ, pin 10, is used by a device to indicate to the controller that the device needs attention. This is the equivalent of an interrupt signal to a microprocessor, and is normally used by a talker to indicate that it has data to transfer, or by a listener which needs data. The REN signal allows any device to be operated either from the IEEE-488 bus (remotely) or locally, as from its own front panel or from a test connector.

Though the IEEE-488 bus is used to a considerable extent in the computer control of electronics systems, few small computers have been equipped with the bus, and for the most popular controller computer, the IBM PC (and its many 'clones'), the IEEE-488 bus interface is fitted as an extra by way of a plug-in card. More specialised microprocessor controllers, however, use the IEEE-488 bus as a standard interface in addition to the serial RS-232.

Serial transfer

The serial transfer of data makes use of only one line (plus a ground return) for data, with the data being transmitted 1 bit at a time. The standard system is known as RS-232, and has been in use for a considerable time. Unfortunately, because the standard is so old and its full implementation is so seldom required nowadays, many manufacturers have made use of serial transfer systems that look like RS-232 and are sometimes referred to as RS-232, but which are not to RS-232 standards. One obvious deviation concerns signal voltages. The original RS-232 system called for voltage levels of around $+15\,V$ and $-15\,V$ to be used for logic 1 and 0 levels respectively. Many modern systems make use of the TTL levels of $+5\,V$ and $0\,V$ in a system that is otherwise RS-232 but which cannot be compatible with a true RS-232 system.

The second complication of RS-232 relates to its two different uses. When RS-232 was originally specified, two types of device were specified as data terminal equipment (DTE) and as data communications equipment (DCE). A DTE device can send out or receive serial signals, and is a terminal in the sense that the signals are not routed elsewhere. A DCE device is a half-way house for signals, like a modem which converts serial data signals into tones for communication over telephone lines or converts received tones into digital signals. The original conception of RS-232 was that a DTE device would always be connected to a DCE device, but with the development of microcomputers and their associated printers it is now more common to need to connect two DTE devices to each other. This requires the connections in the cable to be changed, as

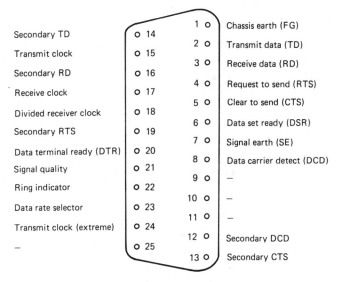

Figure 7.4. The traditional pin assignments for the 25-pin D-connector of an RS-232 link. Many of these reflect the origins of the standard in teleprinter equipment

we shall see. The original specification also stipulated that DTE equipment would use a male connector (plug) and the DCE equipment would use a female connector (socket), but you are likely to find either gender of connector on either type of device nowadays.

The original specification was for a connecting cable of 25 leads, as shown in *Figure 7.4*. Many of these reflect the use of old-fashioned telephone equipment and teleprinters, and very few applications of RS-232 now make use of more than eight lines. The standard connector is the D-type 25 pin, but even in this respect standards are widely ignored and some manufacturers use Din, Amphenol or Cannon connectors. Worse still, some equipment makes use of the full 25-pin systems, but uses the 'spare' pins to carry other signals or even DC supply lines. The moral is that any link that is alleged to be RS-232 must be regarded with suspicion unless the wiring is known from a wiring diagram or from investigation of the connections.

The majority of RS-232 links can make use of eight pins of the 25-pin connector, and these are pins 1 to 7 and 20. *Table 7.3* shows this arrangement on a 25-pin connector; for a lot of equipment, the distinction between chassis earth (pin 1) and signal earth (pin 7) is not necessary. The main data pins are pin 2, the output pin for data transmitted from the DTE to the DCE, and pin 3, the input pin for data from the DCE to the DTE. The use of separate transmit and receive lines means that the serial channel can be used in duplex, allowing transmission and reception of data simultaneously. A full RS-232 implementation allows for two sets of these transmit and receive lines.

Now when a DTE is to be connected to a DCE, the pin 2 of the DTE is connected to pin 2 of the DCE and pin 3 of the DTE is connected to pin 3 of the DCE. The other pins of the DTE are also connected to their corresponding numbers on the DCE. When two DTE devices are connected to each other, however, some links must be crossed. The pin 2 on one DTE must be connected to the pin 3 on one other DTE, and similar cross connection may be needed on handshaking lines. This difference in

Table 7.3.   MOST EQUIPMENT NOWADAYS USES A SUBSET OF THE RS-232 SIGNALS SUCH AS GIVEN HERE, AND EVEN THIS CAN BE REDUCED

Pin	Signal	Action
1	FG	Chassis earth (frame ground)
2	TD	Serial output from DTE to DCE
3	RD	Serial input from DCE to DTE
4	RTs	DTE ready to send data to DCE
5	CTs	DCE ready to accept data from DTE
6	DSR	DCE connected and ready
7	SG	Signal earth
20	DTR	DTE ready to send data

Table 7.4.   THE STANDARD BAUD RATES OF RS-232.
THE SLOWER RATES ARE SELDOM USED, APART FROM THE PRESTEL
USE OF 75 BAUD FOR TRANSMISSION FROM A TERMINAL AND 1200
FOR RECEIVING. NO OTHER DATA COMMUNICATIONS SERVICES USE
THIS SPLIT RATE

50	75	110	150	Slow rates, seldom used
300	600	1200	2400	Used for printers, modems
4800	9600	19200		Fast rate, used for VDU terminals

cabling is indicated by the naming of cables as modem (DTE to DCE) or
non-modem (DTE to DTE), and failure to get a serial link working is very
often due to this very elementary difference.

A serial link can be operated either synchronously (a data bit sent at
each clock pulse) or asynchronously (data sent when ready), and since
practically all modern applications of RS-232 make use of asynchronous
operation, the pin connections for synchronous use can be omitted. For
asynchronous use, each transmitted byte has to be preceded by a start bit
and ended by one or more stop bits. Ten or eleven bits must therefore be
transmitted for each byte of data, and both transmitter and receiver must
use the same number of stop bits. In addition, both transmitter and
receiver must use the same baud rate, the rate of alternation of signal
voltage. *Table 7.4* shows the RS-232 standard baud rates, of which 1200,
2400 and 9600 are the most common. The rates below 300 baud are
hardly used other than by the painfully slow Prestel rate of 75 baud (for
transmitting), and even 300 baud is becoming unusual.

The serially transmitted data will almost certainly use ASCII code (the
older EBDIC code is almost obsolete), which will require only 7 of the 8
data bits that can be sent. Once again, this is not a hard-and-fast rule,
because a serial link to a dot-matrix printer may require all 8 data bits. If
only 7-bit ASCII is needed, then the eighth bit can be used as a parity bit, a
check on the integrity of the data. The parity system can be even or odd. In
the even parity system, the number of logic 1s in the remainder of the byte
is counted, and the parity bit made either 1 or 0 so that the total number of
1s is even. In the odd parity system, the parity bit will be adjusted so as to
make the number of 1s an odd number. At the receiver, the parity can be
checked and an error flagged if the parity is found to be incorrect. This
simple scheme will detect a single-bit error in a byte, but cannot detect
multiple errors nor correct errors. Methods such as cyclic redundancy
checking (CRC) and Hamming or Reed–Solomon codes are needed to
perform such correction and are outwith the scope of this book.

Given that the transmitter and the receiver are set up to the correct
protocols, meaning that the same baud rate, number of stop bits and use
of parity will be identical, we still need hardware methods of handshaking
to ensure that signals are transferred only when both transmitter and

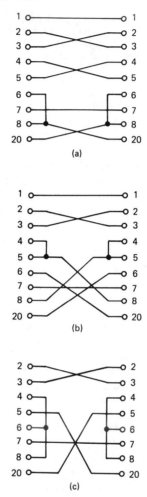

Figure 7.5.   Three common ways of wiring cables or null-modem links. (c) is the recommended linking for the Amstrad PC

receiver are ready. The signals that are used for handshaking are referred to as RTS (ready to send), CTS (clear to send), DSR (data set ready) and DTR (data terminal ready). As usual, you are likely to find that not all four will be used, and only RTS and CTS are common to nearly all equipment.

When a DTE is connected to a DCE (computer to modem, for example), the RTS signal is sent out by the DTE to the DCE to indicate that the DTE has data to transmit. The DCE then responds with the CTS

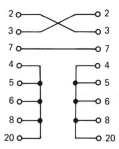

Figure 7.6. An 'auto loopback' connection which dispenses with handshaking. This is useful for testing, and can be used with the XON/XOFF form of software handshaking, or with very low-speed transmissions

signal to indicate that data can be accepted, and data will be sent until the end of the data – note that handshaking is used only at the start and end of each block of data, not between bytes. The handshaking can be correctly implemented for connection of a DTE to DCE simply by using a cable that contains all the necessary strands (a 25-strand cable, for example). For the connection of two DTE devices to each other, a non-modem (or null-modem) cable can be used in which leads are crossed. *Figure 7.5* shows three common connections for such a cable, including the connections for the popular Amstrad PC 1512/PC 1640 machines. An alternative to keeping sets of modem and non-modem cables is the use of 'null-modem' connectors, back-to-back cable connectors which incorporate the reversed leads.

Some more recent developments are useful to note. The RS-423 connection is physically identical to RS-232 but makes use of TTL voltage levels and will tolerate 450 Ω impedance levels. There is a proposed S5/8 standard which makes use of the skeleton of RS-232 in a system that would, if adopted, clear out the present confusion over RS-232, standardise a DIN type of plug and socket for all equipment, and remove the DTE/DCE distinction.

Finally, if you are connecting a serial printer to the serial output of a computer, you can use an 'auto loop back' connection of the form shown in *Figure 7.6*. This makes no use of hardware handshaking and will usually work with no problems at the slower baud rates. If some form of handshaking is still needed, it can be implemented in software by using the XON/XOFF system. This uses the ASCII codes 17 and 19 between computer and printer. Data can be sent to the printer following the ASCII 17 code, and disabled following the ASCII 19. Since these codes are sent over the normal data lines, only the data lines and earth need be connected. The rate of data transfer is slower because of the time that is needed to send the XON/XOFF signals.

Chapter 8

Digital–analogue Conversions

So much signal processing now is digital rather than analogue, and since the display of information is by now much more oriented to digital rather than to analogue methods, the conversion between digital and analogue signals has assumed considerable importance. All conversion from analogue to digital form is based on sampling a waveform and converting the amplitude of each sample into digital signals. The type of digital signal is normally a binary number whose size is proportional to the amplitude of the analogue signal, and the frequency of analogue signal that can be handled depends critically on the rate at which conversion can be achieved. Conversion in the opposite direction has been achieved in the past by methods that used the binary codes of the digital signal to generate analogue voltages which were summed but, as will be explained, this method is often inadequate for precise conversions and has been replaced by other systems, some of which are not quite so new as might be thought at first.

Analogue-to-digital conversion

The conversion of analogue signals into digital form is the essential first step in any system that will use digital methods for counting, display or logic actions. It is often necessary to distinguish between conversion and modulation in this context. Conversion means the processing of an analogue signal into a set of digital signals, and modulation means the change from the original digital signal into a type of digital signal that can be stored or transmitted by an error-free method. The two are very often closely bound up with each other because many forms of conversion are also forms of modulation. In general, if digital signals are to be transmitted over parallel lines only conversion is needed, but if a serial line is to be used, modulation may be necessary in addition to conversion. The more modern methods of analogue-to-digital conversion are based almost exclusively on the modulation system called *pulse code modulation* (PCM), in which each set of digital pulses represents in binary or other number-coded form the amplitude of one sample of the analogue signal, and there is a separate set of digital pulses for each sample of the analogue signal. This system has replaced older methods based on pulse amplitude or frequency, because its output is a stream of digital numbers which can be processed using familiar computing techniques. Such a stream of binary numbers has many advantages over digital signals whose amplitude or frequency is modulated, because variations of amplitude or frequency of a binary signal are not significant provided that they do not cause any confusion between the 0 and 1 bits of each signal. The disadvantage of PCM is that very fast signal processing is required, as will be made clear in the course of this chapter.

The first point to settle about an A–D conversion is how many bits should be used for a number. The use of 8 bits permits the system to distinguish 256 different amplitude levels, and this may be quite sufficient for many applications – the fewer the number of bits in the conversion the faster and more easily the signals can be handled. Much depends on the range of signal amplitudes that need to be coded. If, for example, the signal amplitude range is 90 dB, this corresponds to a signal amplitude range of about 32 000 to 1. Using just 256 steps of signal amplitude would make the size of each step about 123 units, too much of a change. This size of step is sometimes referred to as the *quantum* and the process as *quantisation.* In these terms, 256 steps is too coarse a quantisation for such an amplitude range. If we move to the use of 16 bits, allowing 32 767 steps, we can see that this allows a number of steps that is well matched to the amplitude range of 90 dB or 32 000: 1. This is the level of quantisation that has been used in CD systems. For many instrumentation applications, the range of amplitude levels is much smaller and 8-bit numbers are perfectly adequate. A few applications can make use of 4-bit numbers (up to 16 levels of amplitude), permitting much faster conversion and digital signal handling.

The choice of the number of bits has wider implications, however. Ideally, taking the example of a 90 dB signal-level variation, the amplitude of a signal at each sample would be proportional to a number in the 16-bit range. Inevitably, this will not be exact, and the difference between the actual amplitude and the amplitude that we can encode as a 16-bit number (or whatever is used) represents an error, the *quantisation noise.* The greater the number of bits that are used to encode an amplitude, the lower the quantisation noise will be. What is less obvious is the effect of this difference on low-amplitude signals. For low-amplitude signals, the amount of quantisation noise is virtually proportional to the amplitude of signal, so that when the digital signals are re-converted to analogue the low-amplitude signals appear to be distorted.

The greater the number of digital bits used for encoding, the worse this effect gets, and the cure, by a strange paradox, is to add noise. Adding low-level white noise – noise whose amplitude range is fairly constant over a large frequency range – to very low-amplitude signals helps to break the connection between the quantisation noise and the signal amplitude and so greatly reduces the effects that sound so like distortion. This added noise is called 'dither', and is another very important part of a conversion process for signals with a wide range of amplitude – for many A–D conversions of industrial significance such refinements will be completely unnecessary. The noise level is very low, corresponding to a one-digit number.

Sampling rate

All A–D conversion starts with sampling. Sampling means that the amplitude of an analogue signal is measured and stored in a short interval of time, during which the stored signal level will be converted to digital form, and if sampling is to be used as a part of the process of converting from analogue to digital signals it has to be repeated at regular intervals. The principles of the process are illustrated in *Figure 8.1*, from which you can see that if an analogue waveform is to be converted into digital form with any degree of fidelity, a large number of samples must be taken in the course of one cycle. If too few samples are taken, the digital version of the

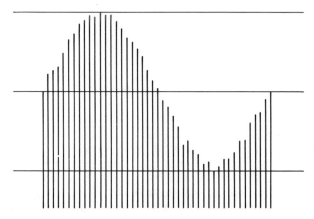

Figure 8.1. Sampling being used on an analogue waveform to create a train of pulses whose amplitude is the amplitude of the analogue signal at the time of sampling

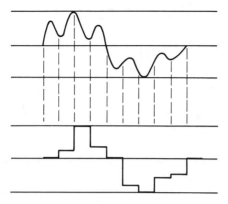

Figure 8.2. The effect of sampling at too low a frequency – the digital levels are a poor representation of the analogue signal

signal (*Figure 8.2*) will look quite unlike the analogue version. On the other hand, if too many samples are taken per cycle, the system will be working with a lot of redundant information, wasting processing time and memory space. Whatever sampling rate is used must be a reasonable compromise between efficiency and precision, and, as it happens, the theory of sampling is by no means new.

In 1948, C.E. Shannon published his classic paper 'A mathematical theory of communications' on which the whole of digital conversion (and much of digital signal communications work) is based. The essence of Shannon's work is that if the sampling rate is twice the highest frequency component in an analogue signal the balance between precision and excessive bandwidth is correctly struck. Note that this pivots around the highest frequency component in a mixture of frequencies – it does not imply that for a 1 kHz sine wave a sampling rate of 2 kHz would be adequate unless your reverse conversion was arranged always to regenerate a sine-wave output. Shannon's theory is concerned with non-sinusoidal waves which can be analysed into a fundamental frequency and a set of harmonics. What this boils down to is that if we look at a typical analogue waveshape (*Figure 8.3*), the highest frequency component is responsible for a small part of the waveform, whose shape looks nothing like a sine-wave and which could equally well be represented by a sawtooth. Sampling the highest *harmonic*, then, at twice the frequency of that harmonic will provide a good digital representation of the overall shape of the complete wave, all other things being equal.

Sampling under these conditions provides a set of pulses whose amplitude is proportional to the amplitude of the wave at each sampled point. A spectrum analyser will then reveal something like the illustration of *Figure 8.4*. This consists of the range of frequencies that were present

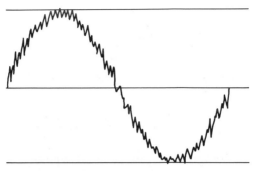

Figure 8.3. A typical complex waveshape in which the highest frequency components form only a very small part of the total amplitude

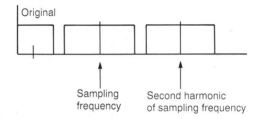

Figure 8.4. A typical spectrum (amplitude plotted against frequency) for sampled signals

in the analogue signal (the fundamentals) plus a set of harmonics centred around the sampling frequency and its harmonics. This is not a problem when the pulse amplitudes are being converted into digital form, but when the pulses are recovered a filter will be needed to separate out the wanted part, which is the lowest range of frequencies, the frequencies of the original signal. The existence of these harmonics makes it even more important to ensure that the sampling rate is high enough. If, for example, the analogue signal had an 18 kHz bandwidth and a sampling rate of 30 kHz were used, the harmonics around the 30 kHz sampling frequency would extend down to $30 - 18 = 12$ kHz and up to $30 + 18 = 48$ kHz, but the lower sideband of this set will overlap the 18 kHz of the original sound, *Figure 8.5*. This is an effect called 'aliasing', meaning that over a range of frequencies in the original range there will be a set of 'aliases' from the lower sidebands of the sampling frequency – the effect is comparable with that of image channel interference in a superhet receiver.

Even if the sampling frequency is made to be twice that of the highest frequency that is normally present in the analogue signal, difficulties still

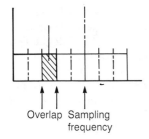

Overlap Sampling
frequency

Figure 8.5. Illustrating aliasing, the over-lap of sidebands from the sampling fre-quency and from its first harmonic

arise because there may be harmonics in the analogue signal at times that extend to higher than half of the sampling frequency. This can be dealt with by using an anti-aliasing filter which is a steep-cut filter that will remove frequencies above the upper limit of the analogue signal's frequency range. If the sampling frequency were too low this filter would need to have an impossibly perfect performance. The sampling frequency must therefore be high enough to permit an effective anti-aliasing filter to be constructed. As before, such methods are needed only for signals whose amplitude and frequency ranges are large, such as audio signals, and in many instrumentation systems using signals at DC or low-frequency AC, such methods are totally unnecessary. With the spread of digital methods, however, the demand for high-performance units is increasing and digital conversion is being used for signals whose frequency range would at one time have been thought to be impossibly high.

Sampling and conversion

The process of sampling involves the use of a sample-and-hold circuit. As the name suggests, this is a circuit in which the amplitude of a waveform is sampled and held in memory while the amplitude size can be converted into digital form. The outline of a sample-and-hold circuit is shown in *Figure 8.6*, with a capacitor representing the holding part of the process. While the switch is closed, the voltage across the capacitor is the

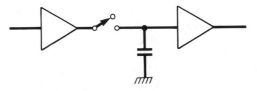

Figure 8.6. The principle of sample-and-hold circuits, with the capacitor perform-ing the holding action when the switch is opened

analogue voltage, maintained by the buffer amplifier which has a low output impedance. No conversion into digital form takes place in this interval. When the switch is opened, the amplitude of the analogue signal at the instant of opening is the stored voltage across the capacitor, and this amplitude controls the output of the second buffer stage. This, in turn, is the signal that will be converted to digital form. The instant of sampling can be very short, but the time that is available for conversion to digital form is the time between sampling pulses. The 'switch' that is shown will invariably be a semiconductor switch, usually a MOSFET type, and the capacitor can be a semiconductor memory, though for a fast sampling rate a capacitor is perfectly acceptable when its only loading is the input impedance of an MOS buffer stage.

The effect of sampling is only to quantise the signal. The signal is still an analogue signal in which the variation of amplitude with time carries the information of the signal, and the change that has come about as a result of sampling is the substitution of a set of pulses of varying amplitude in place of the original continually varying signal. The signal is now an amplitude-modulated set of pulses at the sampling frequency – but this is not a digital signal. The actual conversion from analogue to digital form is the crucial part of the whole encoding process. There is more than one method of achieving this conversion, and not all methods are equally applicable to all uses. There are two main methods and the following is an outline of the problems that each of them presents.

The integrator type of A–D converter is used widely in digital voltmeters. The principle is simple enough, *Figure 8.7*. The central part of the circuit is a comparator, which has two inputs and one output. While one input voltage level remains below the level of the other, the output remains at one logic level, but the logic level at the output switches over when the input levels are reversed. The change at the inputs that is needed to achieve this can be very small, a matter of millivolts, so the action is that the output switches over when the input levels are equal. If one input is the signal being converted, this signal will be held at a steady level (by the sample-and-hold circuit) during the time that is needed for the conversion. A series of clock pulses are applied both to a counter and to an integrating circuit. The output of the integrating circuit is a series of equal steps of voltage, rising by one step at each clock pulse, and this waveform is applied to the other input of the comparator.

When the two inputs are at the same level (or the step waveform input just exceeds the sample input), the comparator switches over and this switch-over action can be used to interrupt the clock pulses, leaving the counter storing the number of pulses that arrived. Suppose, for example, that the steps of voltage were 1 mV, and that the sample voltage was 3.145 volts. With the output of the integrator rising by 1 mV in each step, 3145 steps would be needed to achieve equality and so stop the count, and the count would be of the number 3145 in digital form, a digital

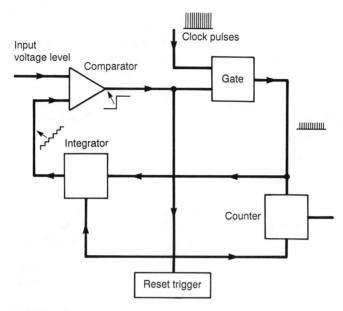

Figure 8.7. The principle of the integrator type of A–D converter as used in digital voltmeters at a slow sampling rate

number that represents the amplitude of the voltage at the sampled input. The switch-over of the comparator can be used to store this number into a register as well as stopping the clock pulses. A master clock pulse can then reset the integrator output to zero and terminate the conversion ready to start again when a new sample has been taken. These numbers are, of course, for illustrative purposes only, but they demonstrate well how a signal level can be converted into a digital number by this method. Note that even with this comparatively crude example, the number of steps is well in excess of the 256 that would be available using an 8-bit digital number. Note also that the time that is needed to achieve the conversion depends on how fast a count rate is used – this will obviously need to be much faster than the sampling rate, since the conversion of each pulse should be completed by the time the next pulse is sampled.

The quality of conversion by this method depends very heavily on how well the integrator performs. An integrator is a form of digital-to-analogue (D–A) converter, so that this method presents the paradox of using a D–A converter as an essential part of the A–D conversion process – rather like the problem of the chicken and the egg. The very simple forms of integrator, like charging a capacitor through a resistor, are unsuitable because their linearity is not good enough. The height of each step must be equal, and in a resistor–capacitor charging system, the height of each step is less than that of the step previous to it. The integrator is therefore very often a full-blown D–A converter in integrated form.

For low-frequency analogue signals there are few problems attached to this method, but for wide ranges of amplitude and frequency the circuits may not be able to cope with the speed which will have to be used. This speed value depends on the time that is available between samples and the number of steps in the conversion. For example, if a time of 20 μs is available to deal with a maximum level of 65 536 steps, then the clock rate for the step pulses must be:

$$\frac{65\,536}{20 \times 10^{-6}}$$

which gives a frequency of 3.8 GHz (**not** MHz), well above the limits of conventional digital equipment. This makes the simple form of integrator conversion impossible for the high sampling rates and large numbers of bits, such as are used for audio-frequency signals. Many of the recent advances in D–A and A–D conversion are due to the research that has been triggered by the use of CD technology and digital tape equipment.

One solution, retaining the integrator type of converter, is to split the action between two converters, each working with part of the voltage. The idea here is that one counter, A, works in the range 0 to 255, and the other, B, in units that are 256 times the steps of the first. The voltage is therefore measured as two 8-bit numbers, each of which requires only 255 steps, so that the counting rate can be considerably lower. The number coded by this method can be written as $(256 \times B) + A$. Since the counters work one after the other, with the smaller range of counter operating only after the coarser step counter has finished (*Figure 8.8*), the total number of encoding steps for a 65 535 size of amplitude is now only $2 \times 255 = 510$, and the step rate is now, still assuming a 20 μs time interval:

$$\frac{510}{20 \times 10^{-6}}$$

a rate of 25.5 MHz. This is well within the range of modern digital circuitry in i.c. form.

Another form of A–D converter is known as the successive approximation type. The outline of the method is shown in *Figure 8.9*, consisting of a serial input parallel output (SIPO) register, a set of latches or PIPO register, a D–A converter and a comparator whose output is used to operate the register. To understand the action, imagine that the output from the D–A converter is zero and therefore less than the input signal at the time of the sample. The resultant of these two inputs to the comparator is to make the output of the comparator high, and the first clock pulse arriving at the SIPO register will switch on the first flip-flop of the PIPO

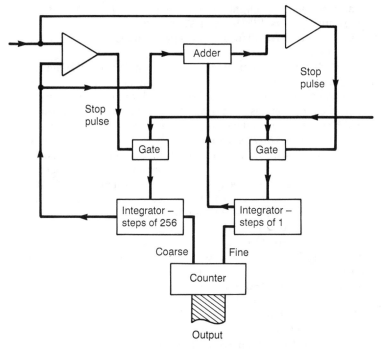

Figure 8.8. Two-stage A–D conversion for higher sampling frequencies. This makes it possible to work at more usable frequencies

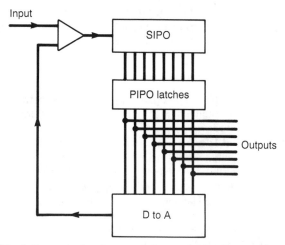

Figure 8.9. A–D conversion by successive approximation, making bit-by-bit comparisons at each level

register, making its output high to match the high input from the comparator. This first PIPO output is connected to the highest bit input of the D–A converter which for a 16-bit register corresponds to 32 767 steps of amplitude.

Now what happens next depends on whether the input signal level is more than or less than the level of output from the D–A converter for this input number. If the input signal is less than this output level, the output of the comparator changes to zero, and this will in turn make the output of the SIPO register zero and the output of the PIPO register zero for this bit. If the input signal is greater than the output from the D–A circuit, the 1 bit remains in the first place of the register. The clock pulse will then pass this pulse down to the next PIPO input – this does not, however, affect the first PIPO input which will remain set at the level it had attained. Another comparison is now made, this time between the input signal and the output of the D–A converter with another input bit. The D–A output will either be greater than or less than the input signal level, and as a result the second bit in the PIPO register will be set to 1 or reset to 0. This second bit represents 16 384 steps if its level is 1, zero if its level is 0. The process is repeated for all 16 stages in the register until the digital number that is connected from the PIPO outputs to the D–A inputs makes the output of the D–A circuit equal to the level of the input signal.

Using this method, only 16 comparisons have to be made in the sampling period for an amplitude of maximum size, giving a maximum time of 1.25 μs for each operation. This time, however, includes the time needed for shifting bits along the registers, and it requires a fast performance from the D–A converter, faster than is easily obtained from most designs. Speed is the problem with most digital circuits, which is why there is a constant effort being made to improve the methods of manufacturing i.c.s, and even to the use of alternative semiconductor materials (like gallium arsenide) that could allow faster operation.

Digital-to-analogue conversion

Conversion from digital to analogue signals must use methods that are suited to the type of digital signal that is being used. If, for example, a simple digital amplitude modulation were being used, or even a system in which the number of 1s were proportional to the amplitude of the signal, the conversion of a digital signal to an analogue signal would amount to little more than smoothing. As it happens, it **is** possible to convert a digital signal into a form that can be smoothed simply, and this is the basis of bitstream methods, noted later in this chapter. For the moment, however, it is more useful to concentrate on the earlier methods.

Smoothing always plays some part in a digital-to-analogue conversion, because a converted signal will consist of a set of steps, as *Figure 8.10* shows, rather than a smooth wave. With a fast sampling rate,

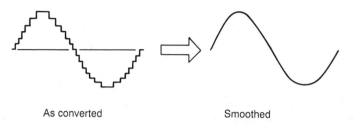

As converted   Smoothed

Figure 8.10. The format of a signal as converted back from digital to analogue form and as subsequently smoothed

however, the steps are closely spaced so that very little smoothing is needed. Even a crude resistor–capacitor smoothing circuit can smooth such a wave quite adequately (though the loss of amplitude may be unacceptable), and the more complex integrators can reproduce the analogue signal very faithfully, often closer to the original than is possible when a purely analogue method of amplification or transmission has been used. The faster the pulse rate of the output of the D–A converter, the easier it is to smooth into an acceptable audio signal.

The pulse code type of digital system does not allow a set of digital signals to be converted into an analogue signal by any simple method, however. Remember that the digital signal consists of a set of digitally coded numbers which represent the amplitude of the analogue signal at each sampling point. The circuits that are used for conversion to analogue will convert each number into a voltage amplitude that is proportional to that number. The nature of the conventional conversion requirement is more easily seen if it is illustrated using a 4-bit number. In such a number (like any other binary number), each 1 bit has a different weighting according to its position in the number, so that 0001 represents one, 0010 is two, 0100 is four and 1000 is eight. The progression is always in steps of two, so that if a register were connected to a voltage generator in the way illustrated for an 8-bit converter in *Figure 8.11*, each 1 bit would

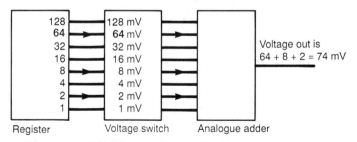

Register   Voltage switch   Analogue adder

Figure 8.11. D–A conversion principles illustrated using a 4-bit number

generate a voltage proportional to its importance in the number. In this way, a 1 in the second place of a number would give, say, 2 mV, and a 1 in the fourth place would give 8 mV, with a 1 in the seventh place giving 64 mV. Adding these voltages would then provide a voltage amplitude proportional to the complete digital number, 74 mV in this simple example. This, in essence, is the basis of all D–A converters.

Reverting to a 4-bit number, *Figure 8.12* shows the basis of a practical method. Four resistors are used in a feedback circuit which makes the output of the buffer amplifier depend on the voltage division ratio. This in turn is determined by the resistor ratio, so that switching in the resistors will give changes in voltage that are proportional to resistor value. In this example, each resistor can be switched into circuit by using an analogue switch, and the switches are in turn controlled by the outputs from the bits of a parallel output register. If the highest-order bit in the register is a 1, the resistor whose value is marked as *R* is switched in, and if the next bit is 1, then resistor 2*R* is switched in and so on.

The result of this is that the voltage from the output of the buffer amplifier (an analogue adder circuit) will be proportional to the size of the digital number. The attraction of this method is its simplicity, which is also its main problem. The resistor-switching type of converter is widely used and very effective – but only for a limited number of bits. The problems that arise are the range of resistance that are required, and the need for quite remarkably precise values for these resistors. Suppose, for example, that the circuit can use a minimum resistance value of around 2k. This is a reasonable value assuming that the amplifier has an output resistance that

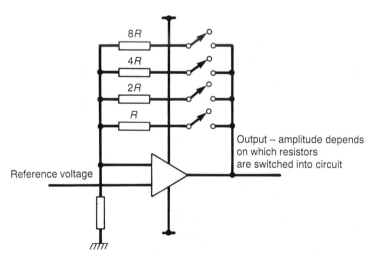

Figure 8.12. A practical D–A converter using a resistor network, illustrated for a 4-bit number

is not negligible and that the resistance values must be large compared with the resistance of the analogue switch when it is on. Now each resistor in the arrangement will have values that rise in steps of two, so that these values are 4k, 8k, 16k . . . all the way to 256k for an 8-bit converter.

Even this is quite a wide range, and the tolerance of the resistor values must also be tight. In an 8-bit system it would be necessary to distinguish between levels that were 1/256th of the full amplitude, so that the tolerance of resistance must be better than one in 256. This is not exactly easy to achieve even if the resistors are precisely made and hand-adjusted, and it becomes a nightmare of difficulty if the resistors are to be made in i.c. form. The problem can be solved by making the resistors in a thick-film network, using computer-controlled equipment to adjust the values, but this, remember, is only for an 8-bit network.

When we consider a 16-bit system, the conversion looks quite impossible. Even if the minimum resistance value is reduced to around 1 kΩ, the maximum will then be of the order of 65 MΩ, and the tolerance becomes of the order of 0.006%. The speed of conversion, however, can be very high, and for some purposes a 16-bit converter of this type would be used, with the resistors in thick-film form. For any mass-produced application, however, this is not really a feasible method. The requirement for close tolerance can be reduced by carrying out the conversion in 4-bit units, because only the highest-order bits require the maximum precision.

Current addition methods

The alternative to adding voltages is the addition of currents. Instead of converting a 16-bit number by adding 16 voltages of weighted values (each worth twice as much as the next lower in the chain), we could consider using 65 535 current sources, and making the switching operate from a decoded binary number.

Once again, it looks easier if we consider a small number, three digits this time, as in *Figure 8.13*. The digital number is 'demultiplexed' in a

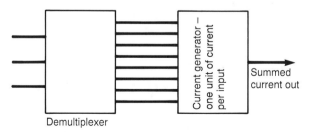

Figure 8.13. A simplified D–A current converter using a demultiplexer to switch equal units of current into an adder

circuit which has eight outputs. If the digital number is 000, then none of these outputs is high; if the digital number is 001 then 1 output is high; if the digital number is 010, then two outputs are high; and if the digital number is 011 then three outputs are high; and so on. If each output from this demultiplexer is used to switch-in an equal amount of current to a circuit, the total current will be proportional to the value of the binary number.

The attraction of this system is that the requirement for precise tolerance is much less, since the bits are not weighted. If one current is on the low side, another is just as likely to be on the high side, and the differences will cancel each other out, something that does not happen if, for example, the one that is low is multiplied by 2 and the one that is high is multiplied by 1024. The effort of constructing 65 535 current supplies, each switched on or off by a number in a register, is not quite so formidable as it would seem, given the current state of i.c. construction.

The main compromise that is usually made is in the number of currents. Instead of switching 65 535 identical current sources in and out of circuit, a lower number is used, and the current values are made to depend on the place value of bits. Suppose, for example, that 16 383 currents are to be used, but that each current could be of 1, 2, 4, or 8 units. The steps of current could be achieved by using resistors whose precision need not be too great to contemplate in i.c. form, and the total number of components has been drastically cut, even allowing for the more difficult conversion from the digital 16-bit number at the input to the switching of the currents at the output. This is the type of D–A converter that is most often employed in fast-working converters.

Another possibility is to use a current analogue of the resistor-switching method used in the voltage step system, so that the currents are weighted in 2:1 steps and only 16 switch stages are needed. This is feasible because current dividers can be manufactured in i.c. form. The

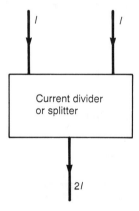

Figure 8.14. The basic unit of a current divider system

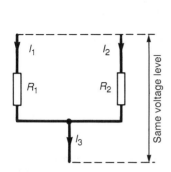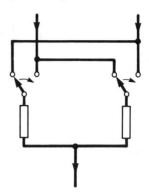

Figure 8.15. The Plassche type of converter using current switching to avoid the need for using very close-tolerance components

principle of a current divider is shown in *Figure 8.14* with currents shown as flowing **into** the terminals at the top rather than out – the distinction between a current adder and a current divider is a matter of which direction of currents you are interested in. To act as a precise divider, however, the input currents must be identical to a very close tolerance, and this cannot be achieved by a resistive network alone.

In this form of converter, first described by Plassche in 1976, the resistor network is used along with current switching. The principle is shown in *Figure 8.15*, and is easier to think of as a current **adder** rather than a divider. The two input currents I_1 and I_2 form the current I_3, and if the resistors R_1 and R_2 are precisely equal, then $I_1 = I_2 = I_{3/2}$, giving the condition of I_1 or I_2 being exactly half of the other current I_3. The snag, of course, is that resistors R_1 and R_2 will not be identical, particularly when this circuit is constructed in i.c. form.

The ingenious remedy is to alter the circuit so that each current is chopped, switched so that it flows alternately in each resistor. For one part of the cycle, I_1 flows through R_1 and I_2 through R_2, and in the other part of the cycle, I_1 flows through R_2 and I_2 through R_1. If the switching is fast enough and some smoothing is carried out, the differences between the resistor values are averaged out, so that the condition for the input currents I_1 and I_2 to be almost exactly half of I_3 can be met even if the resistors are only to about 1% tolerance. Theory shows that the error is also proportional to the accuracy of the clock that controls the switching, and this can be made precise to better than 0.01% with no great difficulty.

Now if a set of these stages is connected up as in *Figure 8.16*, the ratio of currents into the stages follows a 2:1 step, and this can be achieved without a huge number of components and without the requirement for great precision in anything other than a clock signal. In addition, the

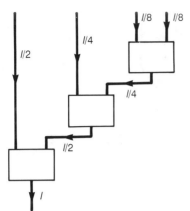

Figure 8.16. Connecting up a set of Plassche stages to form a precise current divider circuit

action of the circuit can be very fast, making it suitable for use in the types of A–D converters that were discussed above. The only component that cannot be included into the i.c. is the smoothing capacitor that is needed to remove the slight current ripple that will be caused by the switching. This can be an external component connected to two pins of the i.c.

The level of conversion is another point that needs to be considered. If we take a voltage conversion as an example, the amplitude of the signal steps will determine whether the peak output is 10 mV, 100 mV, 1 V or whatever. The higher the peak output, the more difficult is the conversion because switching voltages rapidly is, in circuit terms, much simpler when the voltage steps are small, since the stray capacitances have to be charged to lower levels. On the other hand, the lower the level of the conversion, the greater the effect of stray noise and the more amplification will be needed. A conversion level of about 1 volt is ideal, and this is the level that is aimed at in most converters.

Conversion problems

The problems that arise with the conventional type of D–A converter are closely connected with the nature of the digital signal and the ever-present problems of precise current generation. A 16-bit D–A converter requires the use of 16 current sources, and the ratio of a current to the next lower step of current must be exactly 2. For a conventional type of current-generator system in which the current source provides current I, the most significant bit will be switching a current equal to $I/2$ and the least significant bit will be switching a current equal to $I/65\,536$. With all switches open (all bits 0) the current will be zero; with all switches closed (all bits 1) the current will be equal to $I - I/65\,536$. Note that the current can never be equal to I unless the number of bits is infinite; in this example

I/65 536 is the minimum step of current. The value of *I*/2 is taken as being the current corresponding to zero signal, so that zero is being represented by 1000000000000000 (current converters cannot deal with negative currents).

Maintaining the correct ratios between all of these currents is virtually impossible, though many ingenious techniques have been devised. In particular, this scheme of D–A conversion is very susceptible to a form of crossover distortion, when the zero-current level changes to the minimum negative current. In digit form, this is the change from the current *I*/2 (represented by the digital number 1000000000000000) to the value which is the sum of all the current from *I*/4 down to *I*/65 536, corresponding to the digital number 0111111111111111. This step of current ought to be small, equal to *I*/65 536, but if any of the more significant current values are incorrect even to the extent of only 0.05%, the effect on this step amount will be devastating – a rise of seven times the minimum current instead of a fall of the minimum step amount, for example.

This amount of error is virtually unavoidable, and it will cause a crossover error to occur each time the signal passes through the zero level, corresponding to a current of *I*/2. In addition to this crossover problem, all converters suffer from 'glitches', which are transient spikes that occur as the bits change. These are caused by small variations in the time when switches open and close, and they also are most serious when the signal level is at its minimum, because they cannot be masked by the signal level, and the number of bits that change is at its greatest when there is a change from zero (*I*/2 current) to a small negative value. Both of these problems are answered by the adoption of bitstream methods.

Oversampling

The sampling of an analogue signal creates a set of pulses which are still amplitude modulated and which correspond to the analogue signal plus a set of sidebands around the sampling frequency and its harmonics. After the D–A converter has done its work, this is also the signal that will exist at the output and it requires low-pass filtering to reject the higher frequencies.

This, however, calls for the use of filters with a very stringent specification, and such filters have an effect on the wanted part of the frequency range which is by no means pleasant. A simple way out of the problem would be to double the sampling rate. This is not always feasible at the sampling stage since A–D converters are stretched as it is to cope with high-frequency signals. What **can** be done, though, is to add pulses between the output pulses from the A–D converter, *Figure 8.17*. This creates a pulse stream at double the original rate, and so makes the frequency spectrum quite different, *Figure 8.18*. This now makes the

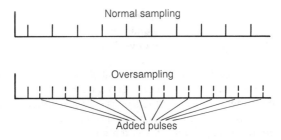

Figure 8.17. *Oversampling is carried out by adding pulses to the received pulses before D–A conversion*

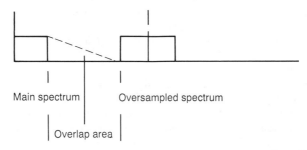

Figure 8.18. *The spectrum of the oversampled waveform, which allows for easier filtering to avoid aliasing*

lowest sideband well above the limit of the wanted signal and easily filtered out. In addition, if the added pulses are midway in amplitude between the original pulses – **interpolated** values – the conversion from pulses to smoothed audio can be considerably smoother, just as if the real sampling rate had been doubled. Oversampling of four times or even more is now quite common.

Bitstream methods

The most significant development in digital conversion in recent years has been the use of bitstream technology in compact disc audio systems. The output of a bitstream converter is not the voltage or current level that is the output of a conventional D–A converter but, as the name suggests, a stream of 0s and 1s in which the ratio of 1s to 0s represents the ratio of positive to negative in the analogue signal. For example, a bitstream of all 1s would represent maximum positive voltage, one of all 0s would represent maximum negative voltage, and one of equal numbers of 1s and 0s alternated would represent zero voltage.

The enormous advantages of bitstream conversion are that only one current generator is required, there are no ladder networks needed to accomplish impossibly precise current division, and the output signal requires no more than a low-pass filter – the bitstream signal is at such a high frequency that a suitable filter is very simple to construct. The disadvantage is that the system must at this point handle very high frequencies – of the order of 11 to 33 MHz for the bitstream converters now being used for audio-frequency signals. The essence of the bitstream converter is that instead of using a large number of levels at a relatively slow repetition rate, it uses very few levels (two) at a much higher repetition rate. The rate of information processing is the same, but it is accomplished in a very different way.

The principles are surprisingly old, based on a scheme called delta modulation which was developed for long-distance telephone links in the 1940s and 1950s. The normal pulse code modulation system of sampling and converting the amplitude of the sampled signal into a binary-coded number has well known drawbacks because of the noise that is generated as a result of the quantisation process. The scheme called delta modulation uses the *difference* between samples, which is converted into binary numbers. If the rate of sampling is very high, the difference between any consecutive pair of samples is very small, and it can be reduced to 1 bit only. A signal of this type can be converted back to analogue by using an integrator and a low-pass filter. Because of the very severe distortion that occurs when a delta modulator is overloaded, the sigma delta system was developed. This uses sampling of the signal amplitude rather than differences, and forms a stream of pulses by a feedback system, feeding the quantised signal back to be mixed with the incoming samples in an integrator. This results in a noise signal which is not white noise spread even over the whole range of frequencies, but coloured noise concentrated more at the higher frequencies. Because of this characteristic, this type of modulator is often termed a noise-shaper. This is the type of technology, used on other digital systems, which has led to bitstream and other forms of D–A converters in which the number of bits in the signal is reduced.

Current bitstream D–A converters consist basically of an oversampling stage followed by the circuit called the *noise-shaper.* The oversampling stage generates pulses at the correct repetition rate, and the noise-shaper uses these pulses to form a stream of 1s and 0s at the frequency determined by the oversampling rate. The important part of all this is the action of the noise-shaper.

Figure 8.19 shows the principles of a noise-shaper, which consists of two adders, a converter, and a delay – the amount of the delay is the time between pulses. If we concentrate on the converter for the moment, this is a circuit which has a very fast response time and which will give an output of 0 for numbers corresponding to less than 0.5 of full output, and an

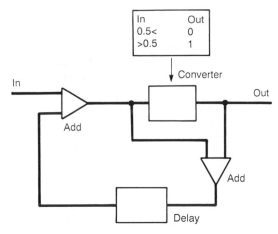

Figure 8.19. The noise-shaper system which is the heart of the bitstream converter

output of 1 for numbers corresponding to more than 0.5 of maximum level. The generation of the bitstream is done rather like the conversion of a denary number into binary, by successive comparisons giving a 0 or 1 and feeding the remainder of a comparison back to be subtracted from the next (identical) input signal. Note that 0 and 1 from the converter corresponds to 0 or I_{max} current values in a conventional D–A circuit rather than the digital 0 and 1 of a number.

Imagine a number at the input causing either a 0 or a 1 signal to be developed. The second adder works with one inverted signal, so that its output is the difference between the original signal amplitude number and the maximum signal (output 1) or minimum signal (output 0). This difference is delayed so that it appears at the input adder while the same oversampled copy of the input is still held there – the amount of delay will fix how many times this action can be carried out on each input. If the error signal is negative it will be subtracted from the input, otherwise it is added, and the result is fed to the converter. This again will result in a max or min output signal and an error difference, and the error difference is once again fed to the input adder to be added to the new input and so to be converted. At each step, the output is a 1 or a 0 at many times the pulse repetition rate of the oversampled signal, and with the ratio of 0s and 1s faithfully following the amplitude corresponding to the input numbers.

Figure 8.20 illustrates the principle. The input number represents an amplitude of 0.7 and this input will cause the converter to issue an output of 1. The difference is -0.3, and when this is added to the next identical input it gives a net input of 0.4 which causes a 0 output with an error of $+0.4$. The diagram shows the successive steps which end when the net

In	Out	Fed back
0.7	1	−0.3
0.7 − 0.3 = 0.4	0	+0.4
0.7 + 0.4 = 1.1	1	+0.1
0.7 + 0.1 = 0.8	1	−0.2
0.7 − 0.2 = 0.5	1	−0.5
0.7 − 0.5 = 0.2	0	+0.2
0.7 + 0.2 = 0.9	1	−0.1
0.7 − 0.1 = 0.6	1	−0.4
0.7 − 0.4 = 0.3	0	+0.3
0.7 + 0.3 = 1	1	0.0 END

Bitstream contains 7 1s and 3 0s; 70% full amplitude or 0.7 V

Figure 8.20. The action of the noise-shaper circuit illustrated for a simple example

input to the converter is 1 so that there is no error signal from the output – this ends the conversion which in this example has taken ten steps – implying that the output frequency will be ten times the input frequency. The output is a stream of seven 1s and three 0s, corresponding to a wave that is 70% of full amplitude, an amplitude of 0.7 of maximum.

Computer plug-in boards

Many computers, particularly the PC type of machine that is almost universally used in business applications, allow for the use of plug-in boards to extend their applications. This, together with suitable software, allows D–A and A–D boards to be used so that the computer becomes part of a control or instrumentation system. A simple example is the use of the computer as a digital voltmeter, indicating voltage either in figures or pictorially. Much more extensive processing can be done even with a simple A–D converter allowing the computer to be used, for example, as a tester for an external PCB by sampling voltages at several different points and determining if they are within tolerance limits.

Some older machines, notably the BBC Microcomputer, incorporated simple D–A converters on board, and later machines of types other than the PC type can also make use of converter boards, either external or (less commonly) internally. Firms who specialise in laboratory and industrial computer applications will supply boards which can be plugged inside computers of the PC type (IBM PC/XT or PC/AT clones or compatibles) so as to make use of the computer for digital signal processing. Such boards include digital input/output cards, A–D and D–A converters. A specialist in such equipment is:

Chipboards Ltd
Almac House
Church Lane
Bisley, Woking
Surrey, GU24 9DR
Tel: (0483) 797959
Fax: (0483) 797702

Chipboards Ltd, along with many other suppliers, advertises in the monthly magazine *Computer Shopper*.

Chapter 9

Computer Aids in Electronics

The availability of low-cost computers has made it possible to carry out many tasks in electronics with very much less effort than was once required. At the time of writing, a good PC computer of the PC/AT type could be bought for as little as £200 (excluding monitor and hard disk), and close scrutiny of the computing magazines will show that some names you might think of as supplying 'basement bargain' computers are, in fact, suppliers of rather overpriced goods. There is not much relationship between price and reliability, and since all PC machines are basically of the same design (and often share the same PCBs) the facilities on the less costly machines are no less than on the high-priced ones. As in so many other aspects of life, it is often better to avoid brand names unless you like paying for a nameplate, and the content of the boxes is more important than their appearance or their nameplates. Two useful sources are listed at the end of this chapter.

In this chapter, three important contributions of the PC computer to electronics are described. These comprise analysis of linear circuits (and there are also programs, not described here, for the analysis of digital circuits), the drafting of printed circuit layouts, and the drawing of circuit diagrams. All three are particularly tedious to carry out by traditional methods, and the use of the computer is an immense saving of time, effort and money, particularly to small firms specialising in one-off equipment. The amateur can also gain considerably from these applications because they provide him/her with methods that were once available only to the largest scale of professional users.

The computer

Any computer described as being PC compatible, XT or AT, can be used for the type of work described here. This includes the Amstrad PC machines but not the earlier PCW machines, and it excludes machines such as the Acorn Archimedes, Atari ST, Apple Macintosh and Commodore Amiga – the use of the letter A is coincidental but a useful way of remembering that these are the incompatible machines, each of which is made by one manufacturer only. Note, however, that some of these other types of machines can be fitted with a PC emulator which allows them to be used with the huge range of software that is available for the PC machines. There are several hundreds of manufacturers turning out PC-compatible machines, and at a very wide range of prices. In general, the older XT design, using the 8088 or 8086 chip, should by now be very cheap, below £200, and the more advanced AT 80286 design can now be obtained at very favourable prices. The lowest at the time of writing was £199+VAT for a 12MHz AT-286 machine with a floppy disk and no monitor (from MicroSurgeons (Isenstein) Ltd). Any machine described as a 286 or 386 machine will be very well suited to this type of work, and if it contains the chip called a maths co-processor the time needed for work such as circuit analysis or drawing will be much shorter. The presence of a maths co-processor, however, is not essential, and the cost would be justified only if you were using the programs for professional purposes.

In general, for your use of the computer you will need at least a high-density floppy disk, preferably the standard 5.25-inch HD type, a monitor which can be colour or (preferably) monochrome, and a printer which can be any type described as Epson compatible. If you intend to use the machine intensively for all the applications described here, particularly for computer-aided drawing (CAD), PCB layouts and other work, you will be advised to add a hard disk of the IDE type – a 40 Mb disk of this type costs just over £100 at the time of writing, and prices were still dropping. One of the considerable advantages of the PC machines is that the scale of production allows important add-ons like hard disk drives to be available at very low prices as compared with other machines.

The use of the later type of PC machines based on the 80386 or 80486 chips is ideal – but if you do not have such a machine it is not necessary to buy one in this class simply to run the circuit analysis or other software. Programs run faster on such machines, but the prices are higher than for the older types. Once again, if you need very considerable power for other purposes, such a machine may be ideal and prices are falling rapidly as the chips become more readily available – see Matmos Ltd for very attractive offers on such machines. The immense advantage of the PC type of machine is that compatibility has been maintained so that programs which ran on the IBM PC of 1981 can still be run on the considerably more advanced machines of today. Users of the older types of machines can also run most modern software (though not all), but often at a low speed. Other types of computers are not only incompatible with the PC type and with each other, but are even incompatible with earlier versions from the same manufacturer.

Linear circuit analysis by computer

The analysis of linear circuits is based on the principles of the effect of resistive and reactive components on the amplitude and phase of a sine wave. For very simple circuits, this can be done either by drawing phasor diagrams to scale or by the use of algebra to express the circuit impedance as $R + jX$ where X is the reactive component and j is the square root of minus one. These methods are comparatively simple but very tedious, and when the circuit is one that is only slightly more complicated than the most basic filter, the amount of work is enormously increased. For many standard circuits, the formulae can be obtained from reference books (such as the splendid ITT *Reference Data for Radio Engineers*, now long out of print), but the amount of manipulation that is required becomes much greater, and in many cases you still have a lot of work to do after working out the results of each formula. The repetitive nature of the work, unless a programmable calculator is used, means that it is very easy to make mistakes.

When the circuit is not a standard one that can be looked up in a reference book, the analysis becomes very much more difficult. It amounts then to combining components, resistive or reactive, in series or in parallel, working out the first two, then combining with the next, and so on until the whole circuit has been covered. The aim is to express the effect of the whole circuit in the format:

$R + jX$

so that the impedance magnitude (Z) is the square root of $R^2 + X^2$ and the phase angle (ϕ) is the angle whose tangent is X/R, *Figure 9.1*. This analysis is long, tedious, and very liable to errors. It requires a good grasp of working with complex numbers (numbers including j) and for a large circuit the effort is very considerable. The working for a simple parallel resistor and capacitor will convince you, if you do not already know, that there is quite a lot of work involved.

Another dimension is added when a circuit contains one or more active components. The gain of an active component converts a passive circuit, whose power gain is always less than unity, into an active circuit which can have a power gain of more than unity over a considerable frequency range (the bandwidth). The gain–bandwidth product of the active device needs to be considered, however, as does the effect of the impedances of the active device.

None of this need be a problem for anyone with access to a

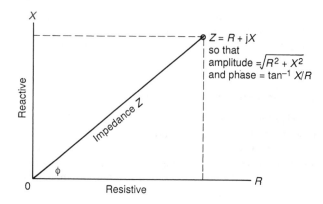

Figure 9.1. Resistance, reactance and impedance in diagrammatic form

PC-compatible computer because there are now several programs which ensure that even for very complex linear circuits, the effort of calculating frequency response can be performed by the computer. This brings linear circuit analysis, once possible only on very large and expensive computers, within the reach of any user, amateur or professional.

There are many linear analysis programs available at the time of writing, but the one that is best suited for the small-scale user, professional or amateur, is called ACIRAN, and is available from the Public Domain Software Library (PDSL) whose address is given at the end of this chapter. ACIRAN has been written by an engineer in the UK and is available in shareware form. For anyone not familiar with this way of distributing software, shareware programs (see the note at the end of this chapter) are not obtainable in the usual way from computer shops but from specialists in Public Domain and Shareware disks. One interesting point is that though most shareware originates in the USA, the ACIRAN program and one other linear analysis program have been written in the UK, so upgrades and support are more readily available.

To obtain the ACIRAN disk, contact the PDSL and ask for disk No. 2609, ACIRAN. At the time of writing, this is version 3.1 of Aciran, but if a newer version becomes available a call to PDSL will confirm the serial number and the version. If you obtain the disk from any other reputable shareware dealer (any one that is a member of the Association of Shareware Professionals) you will also get the current version, but some suppliers may still be stocking the older versions.

Make sure that you get the correct floppy disk size for your computer. Many PC machines use the older type of 5.25-inch floppy disk, but the more recent models use the 3.5-inch type. Each size of disk comes in two capacities, normal or high density, measured as the number of kilobytes

(Kb) of capacity (1 Kb = 1024 stored characters or codes). The low-capacity 5.25-inch disk will store up to 360 Kb, and the high-density disk can store up to 1.2 Mb (1228 Kb). The normal 3.5-inch disk can store up to 720 Kb and the high-density type can store 1.4 Mb (1440 Kb). Note that 1 Mb = 1024 Kb; the use of 1024 arises because this is an exact power of two, 2 to the power of 10, and is better adapted for measuring computer storage than the 1000 that is more familiar to the electronics user. A machine which is fitted with a high-density disk drive can read either type of disk in that size, but the 5.25-inch high-density drives will not write data on to 360 Kb low-density disks.

In addition to its obvious applications to printing tables and graphs of response for a circuit, the use of the Aciran program makes it possible to allow for the effect of input and output impedances and of component tolerances, something that is particularly time consuming if done in the traditional ways. This is, however, the type of information that is particularly needed for small-scale production circuit design, so the use of methods based on the computer is a valuable aid to anyone involved in such work. In addition to the conventional R, C and L passive components, transmission lines can be dealt with, a considerable help to anyone involved in UHF design. As well as the built-in systems for dealing with transistors, FETs and OPamps, there is a current-generator component which can be used to simulate the action of any active components much more closely at high frequencies.

The menus

When Aciran is started, it displays a main menu with headings of File, Modify, Config, Data, Analyse, Results and Graph. The File menu contains the options of New, Load, Save and Quit, allowing you to switch to a new circuit, load in a previous circuit, save the circuit data you are working on, or end the program. The Modify menu allows you to add, change, or delete components in the circuit, name the circuit, change the output point (node, see later) or replace one type of component by another.

The important Config menu determines the form of analysis that will be used, and its first option is Format, allowing the output to be in Polar (amplitude and phase angle) format or Cartesian (resistive and reactive parts). The default is Polar, since this is the format most familiar in linear analysis. The generator and load impedances can also be specified here in the forms of parallel impedances. The default is 100 MΩ resistive for each, so that the circuit will effectively be operating with infinite input and output resistances.

The ReturnLoss option of the Config menu allows for this factor to be calculated for lines, and you can also specify whether or not you want figures for the actual input and output impedance values of the circuit.

Both of these quantities will use the format that you have selected, either polar or Cartesian, and you can make separate runs with different formats selected if, for example, you want to see the impedances in Cartesian format.

The most important selection in this menu is the Sweep, which by default is logarithmic. The alternative is linear, and the option refers to the frequency scale only; the amplitude scale is always in decibels and hence logarithmic. There is also the option to take tolerance into account, with options of on or off. In general, tolerance should be left on only if the effects of tolerance must be investigated – it is usually better and certainly much faster to make the first runs on a circuit with tolerance switched off so as to get an idea of how the circuit behaves. The last item in the Config menu is to switch loudspeaker warnings on or off.

The Data menu consists of two options only, Display or Print. The Display option is the default, so that results will be seen on the screen, a much faster process than printing. Once the settings have been checked and the results look satisfactory, a print run can be made by selecting the other option of this menu.

The Analyse menu can be used only if circuit data has been loaded or typed in to Aciran, and the format of the menu is shown in *Figure 9.2.* You are asked to specify the start frequency and the end frequency for the analysis run. If a logarithmic sweep is being used, the end frequency must be more than ten times the start frequency, but if a linear sweep has been selected in the Config menu, the only requirement is that the end frequency should be greater than the start frequency. You are reminded if you select incorrect values.

You are also asked to specify the number of frequency steps. The greater the number of steps you select (the maximum is 100) the more precise the results will be, but the process of analysis will take longer. When the response is a fairly smooth curve it is not necessary to specify a large number of steps, but for a jagged response a large number of steps will be needed, and it may be necessary to look at a linear sweep over a restricted part of the frequency band to see details of critical parts of the response.

You are also asked to specify the number of tolerance passes. This can be ignored if you have opted in the Config menu not to use tolerance. If you have opted for tolerances this entry is used to specify how many times the circuit will be analysed with different component values. The analysis uses the Monte-Carlo method, with each component being assigned a value at random within its tolerance limits. When all of the tolerance runs are completed, the output shows the upper and lower limits of response. The response curves are always rather irregular if a small number of tolerance runs is used, but the time required for a large number of runs on a complex circuit can be prohibitive unless a very fast computer (with a clock rate of 30 MHz or more) is being used.

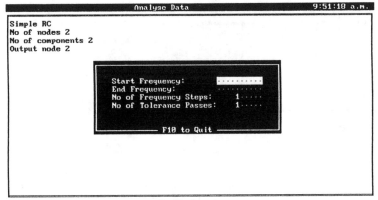

Esc To Cancel

Figure 9.2. The Analyse menu of Aciran as it appears on the screen when a circuit description file has been loaded in

The Results menu allows the table of values (amplitude and phase angle, if you opted for Polar, for frequency response along with time delay, plus any other options) to be displayed on the screen, saved in a disk file for later use, or printed. As usual, display on the screen is a useful first option, and is automatically carried out following analysis, so that this option of the Display menu is used only if you want to re-examine the tables following a graph display. Finally, the Graph menu allows you to show on the screen graphs of the response curves. The scaling of these graphs is performed automatically, so that the numbers used in the scale are often not easy to follow, but the shape of the graphs is usually the more important factor since values can be read from the tables.

The graphical display uses the graphics screen mode of the computer, and provision is made for printing each graph (slowly) on an Epson-compatible printer. Users of laser printers compatible with the Laserjet need to use a subsidiary program to achieve printing on these machines. The most recent version of Aciran allows Laserjet output.

Circuits and nodes

Aciran cannot be used until it has been notified of the circuit components and connections, and since Aciran cannot read a circuit diagram (though if circuit diagrams were prepared with standardised drafting programs such as ORCAD this could be done) it is necessary to use a system for entering the component positions and values. This system depends on identifying and numbering the circuit nodes, as you would when laying the circuit out for construction on PCB or stripboard.

Figure 9.3 shows a typical simple (in terms of components) passive

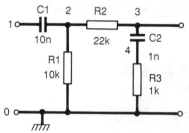

Figure 9.3. A simple circuit for Aciran to analyse with its node points marked. Aciran allows circuit designations such as R1, C2 to be incorporated in a description

circuit, with input and output and a common earth line, with nodes marked by numbers. A node, in this sense, means a point where components join, and this will normally include the input and the output as well as the earth (ground) connection. Each node can be numbered, and the convention followed by Aciran is that the earth (ground) node is always 0 (zero) and the input node is always 1. Other nodes can be numbered as you please, but it is better to take a logical arrangement of node numbers, moving from input to output.

The general rule about nodes is that no node can have more than one number, and nodes are always separated by components. If there is no component between two nodes, one of the nodes must be redundant. If you need to work with nodes connected, use a small resistor value such as 0.33R between the nodes. You can use whatever number you like for the output node, because this will be notified to Aciran, but you **must** use the numbers 0 and 1 for the earth and input nodes respectively. Other linear analysis programs allow you to specify node numbers for all points, but this simply takes longer to set up; since every circuit will have an input and an output there is no reason why these should not be pre-allocated with node numbers.

The total number of nodes in this circuit example is five (numbered 0 to 4). There are also five components, but this is purely coincidental because the number of nodes does not depend in any simple way on the number of components. The point to watch in this illustration is the node numbered 4, because it is easy to overlook this one. Nodes where three or more components join are easy to spot on a well-drawn diagram because of the blob at the junctions, but this type of two-component junction is often less easy to see.

The form of the data can be seen in part in *Figure 9.4*, a view of the Aciran screen when the Data option of the menu has been selected. Small circuits can be checked from this display, but for a complex circuit it is better to print the data out and check against the circuit diagram, which should be numbered to show node positions. Once this set of data items has been entered, component values can be changed, but you cannot

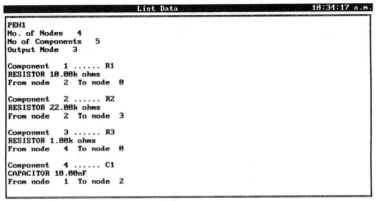

| List Data | | 10:34:17 a.m. |

```
PEH1
No. of Nodes    4
No of Components   5
Output Node    3

Component    1 ...... R1
RESISTOR 10.00k ohms
From node    2  To node  0

Component    2 ...... R2
RESISTOR 22.00k ohms
From node    2  To node  3

Component    3 ...... R3
RESISTOR 1.00k ohms
From node    4  To node  0

Component    4 ...... C1
CAPACITOR 10.00nF
From node    1  To node  2
```

Press Any Key To Continue

Figure 9.4. Part of a data list from Aciran which can be used to check that component values and positions have been entered correctly

change the connections of a component. If you find, for example, that you have mistakenly entered R3 as lying between nodes 3 and 0 and wish to change this to nodes 4 and 0, you have to do this by deleting the component and re-entering R3 with its correct value between the correct nodes. The list will always show that a component has been deleted as a reminder of what you have done.

The form of entry of values follows the standard methods of specifying value, so that a figure by itself is taken to be in fundamental units of ohms, farads or henries. The suffix values of k, M, m, n, p and so on (lower case or capitals except for M and m) are all recognised, and you can enter values in the format 3K3 if you wish.

The analysis of this circuit is started with a wide frequency range and logarithmic response selected. This is always advisable so that the overall picture of the response can be seen, allowing you subsequently to change the frequency limits and sweep type if you want to see more detail. By specifying a large range initially, you ensure that the circuit has no surprises lurking outside the frequency range for which it is intended. This is unlikely in such a simple circuit, but the point about using Aciran is that it allows you to analyse circuits that are far from simple.

The table that appears following analysis takes the form shown in *Figure 9.5* – this is only a part of the table as displayed on the screen. For each frequency in the range, the values of magnitude (amplitude), phase angle and time delay are printed, using units of decibels for amplitude, degrees for phase angle and seconds for time delay. The graph for amplitude is obtained by selecting Graph from the main menu, and appears, *Figure 9.6*, as the first of a set of at least three.

Frequency(Hz)	Magnitude(db)	Phase(Deg)	Time Delay(Sec)
1.000E+02	−24.062	85.256	
1.413E+02	−21.804	83.308	−1.312E−04
1.995E+02	−18.128	80.573	−1.303E−04
2.818E+02	−15.214	76.758	−1.288E−04
3.981E+02	−12.382	71.494	−1.257E−04
5.623E+02	−9.699	64.387	−1.202E−04
7.943E+02	−7.275	55.140	−1.107E−04
1.122E+03	−5.249	43.778	−9.632E−05
1.585E+03	−3.749	30.811	−7.781E−05
2.239E+03	−2.832	17.052	−5.846E−05
3.162E+03	−2.482	3.138	−4.185E−05
4.467E+03	−2.676	−10.654	−2.937E−05
6.310E+03	−3.426	−24.101	−2.027E−05
8.913E+03	−4.758	−36.622	−1.336E−05
1.259E+04	−6.634	−47.318	−8.081E−06
1.778E+04	−8.934	−55.420	−4.333E−06
2.512E+04	−11.504	−60.578	−1.953E−06
3.548E+04	−14.199	−62.766	−5.866E−07
5.012E+04	−16.895	−62.069	1.323E−07

Transmission Results   10:46:55 a.m.

Press Any Key To Continue

Figure 9.5. Part of the table that appears when an Analysis run is completed. This shows the magnitude of gain, phase angle and time delay for the circuit

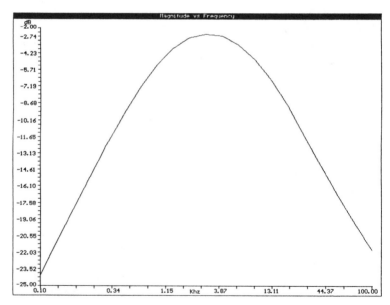

Figure 9.6. The graph of amplitude (magnitude) of gain plotted against frequency for the circuit of Figure 9.3

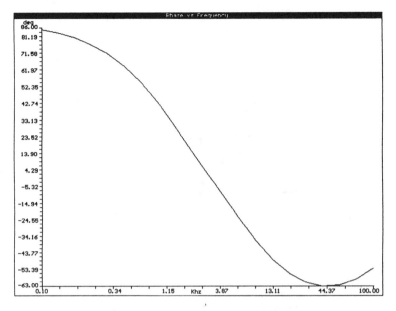

Figure 9.7. The graph of phase angle plotted against frequency for the circuit of Figure 9.3. This shows the considerable changes of phase angle which are often ignored or forgotten by users of such circuits

The other graphs are phase, *Figure 9.7*, and time delay, of which the time delay is usually of least interest for linear circuit work other than for specialised purposes. When option such as return loss and impedance values have been selected, there will be several more graphs of magnitude and phase angle for each of the other quantities covered.

Figure 9.8 shows an active circuit with nodes already numbered. This uses four nodes numbered 0 to 3, and the point to watch here is that Node 0 is applied to both the power supply positive line and to the earth line. This is almost universal in active circuits because the power supply will normally be decoupled to earth and so is at AC earth for all frequencies that will be used. If a circuit contains any other line with is decoupled to earth (a negative supply line, for example, in an OPamp circuit) this also will be numbered as Node 0. Aciran takes no notice of components that are used purely for bias unless they are also significant in the signal path, so that you can analyse circuits without needing to work out the bias, though you have to specify how much current a transistor will pass in no-signal conditions. Another analysis can be performed once bias components have been worked out in order to check that the addition of correct biasing has not materially affected the performance of the circuit.

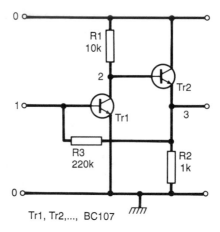

Figure 9.8. An active (transistor) circuit in which input and output impedances are of more interest than frequency response

When Aciran analyses an active circuit, it uses a mathematical description of the active device, and this must be supplied. You can opt to supply the details for yourself, but the more useful alternative is to use a ready-made device file. The form of the description is as a short file consisting of words and numbers (a MODEL file) and V.3.0 of Aciran supplies some of these files. The form of the file is simple, and you can readily add new devices for yourself, provided that you can find the necessary parameters.

The circuit above contains no capacitors, so that its frequency response is irrelevant. It would be possible to add capacitors to represent the effect of strays if frequency response was the important factor, but in this example the input and output impedances are more important. *Figure 9.9* shows part of the table of impedance results for this circuit, with input impedance magnitude and phase followed by output impedance magnitude and phase for each frequency value in the table. The values indicate a fairly constant 300 ohm input and 7 ohm output resistance value over the range, which was 100 Hz to 100 kHz.

The amplitude graph for this circuit, *Figure 9.10*, shows that the response has dropped by about 0.5 dB at 100 kHz, and this must be due to the characteristics of the BC107 transistors rather than to the effect of capacitances. The graph of input impedance magnitude, *Figure 9.11*, shows the low-frequency value of just over 337 ohms rising to around 357 ohms at the higher frequency limits, and the graph of output impedance, *Figure 9.12*, shows a rise from 6.9 ohms to 7.7 ohms which would probably not be particularly significant. Consider how much effort would be needed to obtain these figures by manual calculation, and you

Impedance Results				11:03:08 a.m.
(Hz)	(Mag)	(Pha)	(Mag)	(Pha)
1.000E+02	3.377E+02	-0.000	6.922E+00	0.003
1.148E+02	3.377E+02	-0.000	6.922E+00	0.004
1.318E+02	3.377E+02	-0.000	6.922E+00	0.004
1.514E+02	3.377E+02	-0.000	6.923E+00	0.005
1.738E+02	3.378E+02	-0.000	6.923E+00	0.006
1.995E+02	3.378E+02	-0.000	6.923E+00	0.006
2.291E+02	3.378E+02	-0.000	6.923E+00	0.007
2.630E+02	3.378E+02	-0.000	6.924E+00	0.008
3.020E+02	3.378E+02	-0.001	6.924E+00	0.010
3.467E+02	3.378E+02	-0.001	6.924E+00	0.011
3.981E+02	3.378E+02	-0.001	6.925E+00	0.013
4.571E+02	3.378E+02	-0.001	6.925E+00	0.015
5.248E+02	3.378E+02	-0.001	6.926E+00	0.017
6.026E+02	3.378E+02	-0.001	6.926E+00	0.019
6.918E+02	3.379E+02	-0.001	6.927E+00	0.022
7.943E+02	3.379E+02	-0.001	6.928E+00	0.025
9.120E+02	3.379E+02	-0.002	6.929E+00	0.029
1.047E+03	3.379E+02	-0.002	6.930E+00	0.033
1.202E+03	3.380E+02	-0.002	6.931E+00	0.038

Press Any Key To Continue

Figure 9.9. Part of the impedance versus frequency plot for the OP-amp circuit of Figure 9.8 showing magnitude and phase for input (left) and output (right)

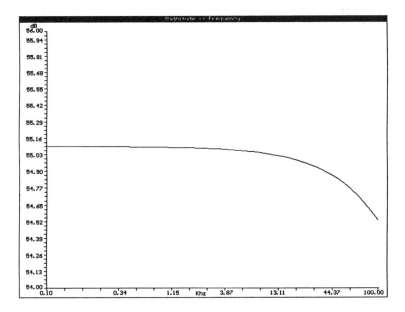

Figure 9.10. The graph for amplitude plotted against frequency and showing a fairly constant gain that starts to fall off at 100 kHz

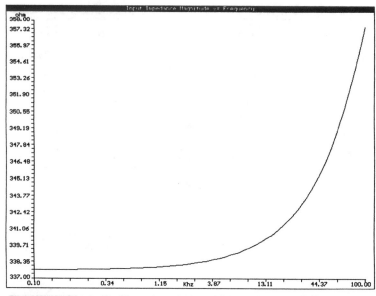

Figure 9.11. The graph of input impedance magnitude for the circuit of Figure 9.8 shows the way that this quantity rises steeply as the frequency approaches 100 kHz

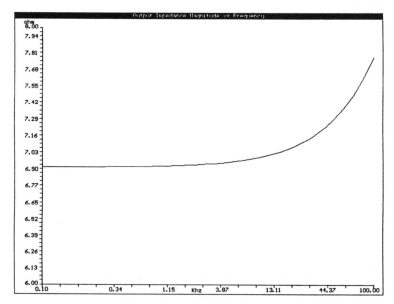

Figure 9.12. The graph of output impedance magnitude for the circuit of Figure 9.8, showing the less significant rise as the frequency approaches 100 kHz

can see what an immense step forward the Aciran program brings to small-scale circuit designers.

Figure 9.13 shows a linear i.c. circuit which uses six nodes numbered 0 to 5. The positive and negative supply lines are regarded as part of Node 0 like the earth line, and you need to remember that the connection of the non-inverting input is another node, since it is connected to a component, the resistor. The node marked as 3 is also one of the type that is often forgotten.

This time, the circuit will be assigned tolerances, using 10% for the resistors and capacitors, and the tolerance setting of the Config menu will be turned on. In the Analysis menu, 100 tolerance passes will be specified, since this is a simple circuit, and the analysis will take several minutes because of this number of passes. The results will follow the familiar pattern of output magnitude and phase, but with the effects of the tolerances displayed this time. When a circuit is being tried out, always find how long a no-tolerance run takes, because this will be a guide to the time needed for a tolerance run. For elaborate circuits with a large number of tolerance runs, the computer can be kept working overnight, because even a fast machine can take a long time over so many analysed points. Only the fastest machines that are fitted with the maths co-processor chip will return the results of multiple tolerance runs in a reasonable time.

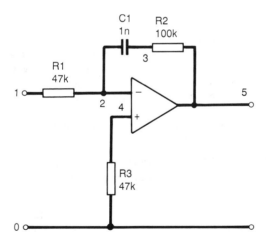

Figure 9.13. A linear OP-amp circuit – note that bias components can be ignored, so that OP-amp circuits are particularly easy to work with

Frequency(Hz)	Magnitude(db)	Phase(Deg)	Time Delay(Sec)
	Transmission Results		11:41:05 a.m.
1.000E+02	32.323	94.151	
1.148E+02	31.149	94.733	2.581E-04
1.318E+02	29.818	95.581	3.100E-04
1.514E+02	28.653	96.252	2.789E-04
1.738E+02	27.551	96.983	3.125E-04
1.995E+02	26.114	98.326	2.495E-04
2.291E+02	25.204	99.601	2.963E-04
2.630E+02	23.755	101.000	3.368E-04
3.020E+02	22.631	102.333	3.035E-04
3.467E+02	21.703	104.057	2.771E-04
3.981E+02	20.401	106.353	2.450E-04
4.571E+02	19.310	108.706	3.161E-04
5.248E+02	18.054	111.160	2.818E-04
6.026E+02	17.004	114.333	2.613E-04
6.918E+02	16.129	117.177	2.304E-04
7.943E+02	14.917	119.462	2.478E-04
9.120E+02	13.845	123.647	2.158E-04
1.047E+03	13.441	128.110	2.410E-04
1.202E+03	12.614	131.256	1.953E-04

Press Any Key To Continue

Figure 9.14. A portion of the first table from a set of tolerance runs – two tables are printed, one for each limit of tolerance

Figure 9.14 shows a sample of the form of the table that is obtained when tolerance runs have been specified. When a tolerance analysis has taken a considerable time, such tables should be saved to a disk file because they represent a considerable investment in time and effort. The appearance of the table is unchanged because two tables are provided, one for each limit of tolerance. The graphing of the tolerance results is less simple – it is very difficult to persuade Aciran to draw the true tolerance graph, *Figure 9.15*, and you will normally see only the single trace that is presented when no tolerances have been specified. The method that is required to show the double-line tolerance graph when only the single-line version is shown is to return to the Analysis menu, but to escape from it by pressing the Esc key (without entering any data) and then opt for Graph.

Aciran keeps files for many of the popular linear i.c.s, so that if your circuit uses one of the set:

LM124	MC1558	NE530	NE538
NE5512	NE5532	NE5534	TL084
UA741	UA747		

it can be accommodated. There is also a STANDARD option for any OPamp whose characteristics conform to the pattern of:

Input impedance	100 MΩ
Output impedance	600R
Open-loop gain	100 dB
Gain × bandwidth	100 MHz
Tolerance of open-loop gain	50%

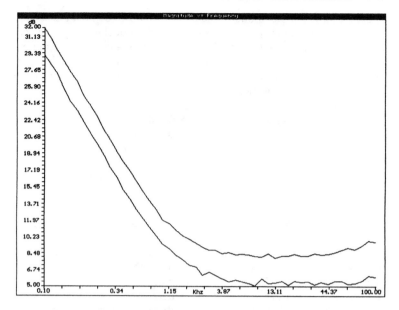

Figure 9.15. The graph of magnitude plotted against frequency for a set of tolerance runs. The graph outlines are more jagged because of the Monte-Carlo method used, and it is not easy to obtain the correct graph, see the text

As for transistors, you can add i.c.s to the existing MODELS set for yourself, typing in the values from a databook. The MODELS list also includes MOSFETs and the same remarks apply to these. The main snag is that the parameters which need to be known for transistors and FETs are not so easy to obtain. Most collections of transistor and FET data contain only the DC parameters, along with h_{fe} and F_T. Aciran demands figures for the parameters h_{ie} and h_{oe}, together with the tolerance of h_{fe} values. These figures can be obtained from the manufacturers' manuals, such as the *Philips Technical Handbooks*. It is also possible to obtain disks of data, often at much lower prices than the manuals, from sources such as PDSL.

The same comments apply to FETs for which details of g_m, c_{gs} and c_{gd} are needed, along with the percentage tolerance of g_m. The figures for OPamps are the open-loop gain and tolerance of open-loop gain, along with the gain–bandwidth figure (not always quoted), and the input and output impedance figures. Though it is generally easier to find the parameters for OPamps, not all collected databooks include the gain–bandwidth figure, and virtually none provide the figures needed for the FETs.

The immense value of ACIRAN is that it allows the effects of modification of a circuit to be tried out without the need to construct the circuit. Changes in component tolerances, values of damping resistors, altered transistor bias current, changes of input/output impedance can all be tested on the computer model in a fraction of the time that would be needed to construct and test the circuit, and without the complications of allowing for the effects of measuring instruments. If semiconductors need to be used at frequencies close to their limits, a more elaborate mathematical model can be used to make the analysis more exact. Since each circuit can be saved in the form of a disk file, it is possible to build up a library of circuits, allowing each to be used as the basis for other work (for example, an elaborate filter circuit can have its component values changed so as to cover a different frequency range).

PCB layouts

Programs for PCB layout have in the past been prohibitively expensive for the small-scale user, though the Hi-Wire package (from Riva, Tel: 0420 226666) selling at £695 is suitable for the smaller firm. If your needs are less extensive or your budget more limited, the PDSL can supply a program called PC TRACE (formerly known as PC Route). This is obtainable on a shareware basis for the usual copying fee of around £4.00 and can be licensed for around $75 payable direct to the author, Doug Ehlers, in the USA. The description that follows is for the use of PC TRACE, because the principles that it uses are common to other PCB layout programs, some of which are very costly indeed and run only on large computers.

PC TRACE is the first PCB drafting program to be offered as shareware, and in the latest version new features have been added such as variable pad and trace sizes, computer mouse support, and support for printers of the Laserjet, PostScript, HP Deskjet, and Epson MX-80 types. It features autorouting, graphic interactive routing, board layout functions, flexible pinouts for devices, and the use of up to 300 components and 1800 connections.

When PC TRACE is run, you can either make use of a file of connections that has already been prepared or notify components and connections directly. Like many programs, PC TRACE depends on a main menu, *Figure 9.16*, from which you can select your choices, and in this main menu you would normally start by selecting *Define/Edit board*, and then *Define board dimensions*. The maximum board size supported by PC TRACE is 11 × 8 inches and the unit of measurement is 0.05 inches. This means that for a 4-inch square board you need to enter dimensions of 80 × 80. You are allowed to change board sizes at any time during the design process, but if you make the board smaller, any chips that were placed outside the new border will have to be replaced. Esc will back up to the previous prompt, or jump out of the routine leaving the board size unchanged.

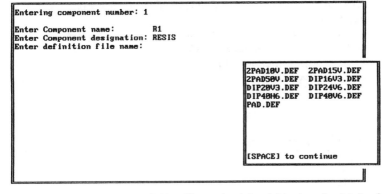

```
        Main Menu

  1) Input/Output menu
  2) Print out current data
  3) Define/Edit board
  4) Route board as per memory
  5) Interactive router
  6) Show board statistics
  7) Exit program

  Choice: _
```

Figure 9.16. The main menu of PC TRACE as it appears on the screen. Select by typing a number and then pressing the ENTER key

```
Entering component number: 1

Enter Component name:        R1
Enter Component designation: RESIS
Enter definition file name:
                                   2PAD10V.DEF  2PAD15V.DEF
                                   2PAD50V.DEF  DIP16V3.DEF
                                   DIP20V3.DEF  DIP24V6.DEF
                                   DIP40H6.DEF  DIP40V6.DEF
                                   PAD.DEF



                                   [SPACE] to continue
```

Figure 9.17. Referring to a definition file, on the right of this view, for details of shape and size of components

Once the board size has been determined you can select entry of new components from the same Board menu. The program will prompt you with questions for each component such as the name of the chip or component, type of component, definition file (for shape and size, see *Figure 9.17*), position on board (such as left side of board), pad size. An entry of N for a component will return you to the board definition menu, and you can check the components list by using the Print/List Components option from the Print menu, *Figure 9.18*. The next step is to enter some connections to be made between the chips by selecting the *Connection entry* section. When this has been done, the set can be

Name	Desig.	Def. File	Pincount
8052 uP	IC1	DIP40V6.DEF	40
741s373	IC2	DIP20V3.DEF	20
2716	IC3	DIP24V6.DEF	24
100k Resistor	R1	2PAD50V.DEF	2
10k Resistor	R2	2PAD50V.DEF	2

[SPACE] to continue

Figure 9.18. The Print menu listing of components which can be used to check that all the components on the board have been entered into PC TRACE

checked by using the Print/List Connections option from the Print menu, *Figure 9.19.*

All traces are assumed to be 0.02 inches wide unless you change this default. You will be asked for each connection if it is a priority connection and at this stage the normal answer will be N. Connections can be entered until you answer N to the question *Enter another connection.* The connections are entered in the form:

Source Device Pin Destination Device Pin

so that typical entries would be of the form:

IC3 2 IC1 3
IC2 6 R1 1

with a final N used to return to the main menu.

The next step is to perform the route action on this board by selecting the *Route board as per memory* option. You will be prompted at this point to select single-sided routing or double-sided routing, and then to press the Esc key. When autorouting is completed the program will display a list of any uncompleted connections to the printer or screen, whichever you wish, and you are returned to the main menu.

The results can be inspected (but need not be) by selecting item 5 from the menu, allowing you to see a form of representation of the board on the screen – it is usually necessary to use the cursor keys or the mouse to see all of the board plan, which is magnified on the screen showing component positions (not in a graphical form) but not connections. If too many connections remain unmade on a single-sided board you can

Device 1	Pin #	Device 2	Pin#
IC1	2	IC3	2
IC1	3	IC2	16
IC1	21	IC1	15
IC1	14	IC2	9
IC2	8	IC3	17
IC2	8	IC3	8
IC2	9	IC3	9
IC2	18	IC3	18
IC3	22	IC2	6
R1	1	IC2	4
R1	2	IC1	31
R2	1	IC1	8
R2	2	IC3	6

[SPACE] to continue

Figure 9.19. The Print menu listing of connections, allowing you to check that all connections have been entered

re-route the board using double-sided methods. The program will then route double-sided and again give a list of incomplete connections if there are any. When this has been done, viewing the board on the screen allows the use of the S key to change the sides of the board that is being viewed, and there are several other options. In general, this Interactive Routing menu item can also be used for routing selectively, controlling paths and trace widths rather than relying on the automatic process triggered by item 4.

When the routing is complete you can select *Statistics* to show how good the design is. The *Equivalent Integrated Circuit Count* is the total number of pins divided by 16, and *Board Density* is the amount of total space that each chip has to occupy, calculated by taking the area of the board and dividing by the Equivalent IC count. The *Total Trace Length* gives the total length of all the traces on the board. The smaller this number, the better the design. The data must now be saved on disk using the *Save Data* option. This will save a complete description of the design layout, component list, and connection list. You will be prompted for a filename to use.

Finally, the PCB design can be printed out, using a separate program section called RPRINT. When this program starts you will be prompted for the printer type and then for the board name. The main control panel will then appear on the screen. You can select another filename at this point if you want to, and then make the selections that are needed to print out the board details. The first selection is the board side to print, and single-sided boards are always printed on the solder side. You can opt to mirror the board left to right if you want to, or switch between printing a negative or a positive of the actual board.

Laser printers and 24-pin dot-matrix printers can produce the images at actual board size; other printers use double-size printing. Once these set-up steps have been taken the printing of the board will be carried out. You can also opt to print a component list, sorted by name or designation, or to print connections, resulting in a list of all the connections for the board, again sorted by the source or destination designation. Another useful option is to print a silkscreen representation. This is not the same scale as the actual board (unless you are using a PostScript type of laser printer) but it is useful for remembering where components are to be placed. Printing can take a considerable time, even for a simple board, but this has to be compared with the much longer time required for a manual design. *Figure 9.20* shows a simple layout, the example that is provided along with the PC TRACE files.

The advantage of using a program of this type is that design is done on-screen, with no need to produce any printed work until you are satisfied that the design is correct. Even when a design has been printed on paper, you can still check the work and amend the design before starting the production of boards. Because each type of component can be held as a disk file there is no delay in specifying any standard

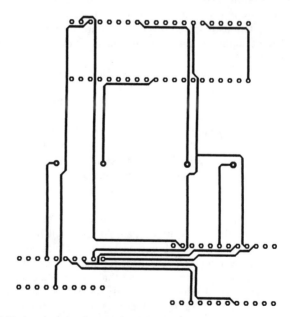

Figure 9.20. An example of a very simple layout printed full size from PC TRACE on a Laserjet printer

component, and if you need to use components that are non-standard or not catered for in the standard set you can enter data to make your own definition files – components can be entered as vertical or horizontal. The weakness of the program is that you cannot see what you are about to print – the picture produced by the Interactive Routing option is not one that gives much real information. In addition, the files that PC TRACE produces are very large even if only a few components are used.

Circuit diagrams

The problems of drawing good clear circuit diagrams have haunted electronics engineers for many years, and can now be solved by the use of any of the low-cost CAD (Computer-aided Design) software packages that are now available. One of the most suitable for the purpose is AutoSketch, a commercial product currently selling for around £100 in its version 3.0. The particular advantage in using AutoSketch as compared with other CAD packages in the same price range is that it is compatible with the AUTOCAD program that is used by large firms (with a price tag to match). An option of particular interest to amateur enthusiasts is EEDRAW from the PDSL (see the end of this chapter) which allows electronics drawings to be made using an Epson printer and VGA screen type (but with few other options).

Using AutoSketch you can produce drawings of professional quality on paper as large as your printer or plotter can handle. Your drawings can include text such as labels, headings and the symbols of mathematics and other specialised applications. You also need a video screen system capable of displaying graphics. AutoSketch can be used with any of the standard systems such as CGA, EGA, VGA or Hercules, but the systems which offer higher resolution, such as Hercules and VGA, are easier to work with. You must have a mouse connected to the computer and suitable software installed. You also need a hard disk and the use of a printer or plotter. Work of good quality can be produced using a dot-matrix printer, but much better results can be produced using a laser printer, inkjet printer, or a pen-plotter such as the Hewlett-Packard, Roland or the remarkably inexpensive ACS-APT type. For work that involves large drawings, it is an advantage to fit an expanded (not **extended**) memory board to your computer. AutoSketch 3 can use up to 2 Mb of such memory. For fast drawing actions, it is an advantage to fit a maths co-processor chip. If your output is to a laser printer it is sometimes an advantage to specify a resolution of 150 dots per inch rather than the default 300, because this makes lines thicker and allows dotted lines to be seen on the printout.

Drawings made using AutoSketch can show as much detail as your **printer** can deal with. Using AutoSketch with a laser or inkjet printer allows you to use up to 300 dots per inch, and with a plotter even finer details can be achieved. AutoSketch must be configured (adjusted) for the type of mouse, type of screen and type of plotter or printer that you use. Configuration is done simply by selecting numbers from each of several lists when the program is first used.

AutoSketch deals with dimensions as numbers when you make a drawing, with no reference to units (feet, inches, metres, millimetres) until you come to print or plot the drawing. You can specify the largest range of dimensions on the normal drawing screen, but this does not restrict you to remaining within these limits. For electronics drawings it is best to use the paper dimensions.

The screen that appears when you run AutoSketch (after configuring) is the Drawing Screen, *Figure 9.21*, which contains the Menu bar, scroll bars, and the arrow pointer. The arrow is used for selection and for drawing, with the scroll bars used to perform the equivalent action of sliding a sheet of paper around on the drawing board. All of your drawing work is done on the screen display, but there are several options about the information that is seen on the screen. For example, the grid of fine dots on the screen can be turned on or off using the *Assist* menu, and the spacing of items is determined from the *Settings* menu.

Some useful aids are included in the Assist menu. Ortho allows only horizontal and vertical lines to be drawn, no matter how the mouse is

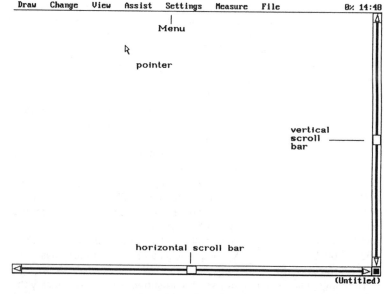

Figure 9.21. The main drawing screen of AutoSketch as it appears when you start the program, showing the menu, selection arrow and scroll bars

moved. This is useful for electronics diagrams in which lines representing connections are either horizontal or vertical. The grid is a more generally useful guide to positioning the mouse, displaying a grid of faint dots on the screen. The spacing between the dots can be altered to a value that suits the drawing size; 1.0 mm is often useful for electronics work. The Grid is a guide, showing more clearly how far you have moved the cursor, and allowing you to move parts of a drawing into their correct positions, or draw them to correct dimensions.

Selecting *Attach* from the *Assist* menu allows you to place either end of a line precisely, something that is not possible simply by moving the mouse. It needs to be switched off if you want to draw lines to or from points other than the ends or middle of another line, for example.

Snaps are used to enforce precision of placing the cursor. When *Snaps* are selected from the *Assist* menu, the arrow cursor is replaced by a cross marker. This marker can be moved only to snap points, which, if you are working with Grid on, will be the same as the grid size so that your cursor will snap to each grid point. When you alter the grid size, however, this does not alter the Snap size. You will need to switch Snaps off if you want, for example, to start a line from the rim of a circle, because the use of Snaps will make this impossible except where the circle coincides with a snap point.

Your first requirement is to create or buy a set of electronics symbols. If you create your own, you need to work with care to ensure that the size of each component is suitable and that its terminals are on snap points, something that is not always easy for some shapes like transistors and inductors. Unless you have time on your hands, it is often better to buy a set of these sub-drawings, called *Parts*. For each shape, whether a small part or a complete drawing, you can rotate, mirror-image and scale the drawings.

An object that is needed in many drawings can be saved as a Part, with some point of the object designated as the *Part Base*, a point for manipulating the object. The object can be drawn alone, with nothing else on the screen, and saved in the usual way, or it can be a portion of a larger drawing and saved using the *Part Clip action*. When the Part is needed, the *Part* option from the *Draw* menu is selected. This gives a list of all files, with small image (icons) to make selection easier. When the Part is loaded it can be moved to any part of a new drawing.

A simple way of creating a collection of Electronics Parts is to draw a circuit with as many types of components as possible, and save each component as a Clip Part. Even better is to collect a number together and save them as a file which can be recalled when you want to create a circuit diagram, *Figure 9.22*. The illustration has been taken from the screen, and the printed version of these components is very much better than the screen version because of the limited resolution of the screen. If you use names such as ER, EC, EL, ET1, ET2 and so on, prefixing each Part name

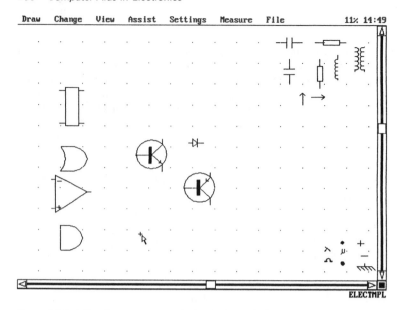

Figure 9.22. A file of component symbols which can be kept unaltered and used as a template for circuit diagrams, erasing from the screen view symbols which are not needed and copying others

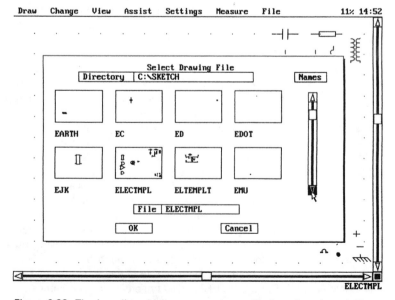

Figure 9.23. The Icon list which appears when a file is to be selected. These ...) views make it easy to select the c(...

with the same letter, all of these parts will be arranged together in the Icon list, *Figure 9.23.*

Any portion of a drawing can be selected so that it can be moved, copied, stretched, or have lines selectively broken. Move and Copy both make a new copy of an object at another position, but when Move has been selected, the original is deleted. This allows circuits to be built up by copying components from one place to another, as shown in *Figure 9.24*, and then drawing in the connecting lines. The use of Snaps will ensure that all components are perfectly lined up, *Figure 9.25*, and then the final work consists of copying over the round blobs that mark junctions in the circuit.

Zoomed views, such as *Figure 9.26*, are used in AutoSketch to use the whole screen for a view of a drawing, or part of a drawing, at a different scale. This can be used to make the detail of part of a drawing more precise, or to make a small drawing fill the screen, or to make a drawing which is larger than the original limits fit into the screen space. This can be very useful if detail needs to be worked on, and it can be used in the opposite direction if the size of a drawing turns out to be larger than you expected.

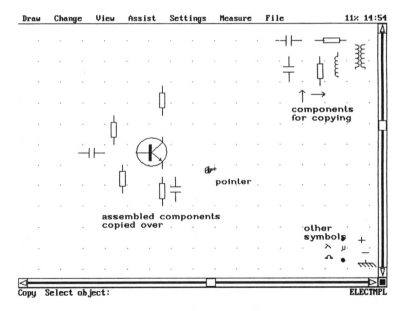

Figure 9.24. AutoSketch in action, copying component symbols from the template file to use in a circuit. Note the grid dots which are used as a guide, and the use of (invisible) snap settings to place symbols precisely

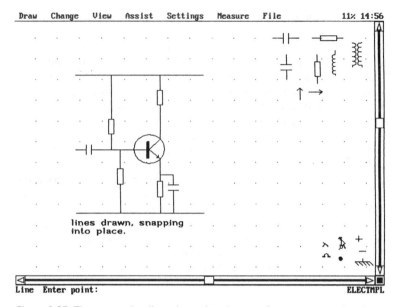

Figure 9.25. The connecting lines drawn in – the use of snaps ensures that these lines meet up precisely with lines or points on the components

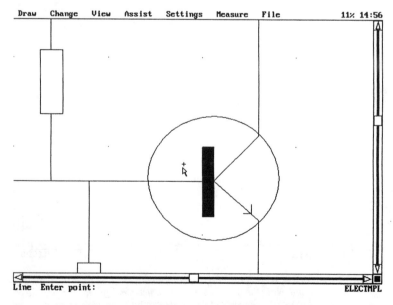

Figure 9.26. Using a zoomed view to attend to detail of a drawing. Drawings can be zoomed almost indefinitely with no loss of detail

Text can be added to any drawing, and the size of text characters is related to the drawing unit – this will almost always need to be changed. There are several sets of characters (or **fonts**); the default font provides characters that are drawn with simple straight-line strokes.

Because each drawing that is created by AutoSketch can be saved as a disk file, it is often possible to make one drawing the basis of another, saving a considerable amount of work, and saving the new drawing under a different filename. Since no drawing need be committed to paper until it is needed, you can adjust as much as you like until the drawing is as you want it, something that is impossible when you work directly on to paper. In addition, where drawings make use of repetitive features (sets of i.c. pins, parallel lines of a bus, etc.) the multiple-copy actions of AutoSketch make the creation of such shapes very fast and easy.

The Public Domain Software Library

The PDSL exists to supply disks of programs that are virtually free for inspection, and the only cost to the user is the cost of copying the disks. PDSL can supply on a range of disk formats, and in some cases is virtually the only source of software for some exotic machines. All of the programs are either public domain or shareware. Documentation for each program is included as a disk file, usually with the DOC extension. There are many other suppliers of the same software items, but PDSL was among the first and provides a more full description of the contents of its disks. Like all other reputable PD and shareware dealers, it also warrants, as far as is possible, that none of its software is pirated commercial software. PDSL is a member of the Association of Shareware Professionals.

A public domain program is one for which all copyright has been waived, so that it can be copied freely, distributed freely, and used without charge other than copying costs. Public domain programs arose in the USA because universities and colleges are required by law to make public property any software that they develop. This became a rich source of some excellent programs, and the range of PD software was extended by programmers working for enjoyment, or whose ideas had not been accepted commercially.

Shareware is a later concept, in which the author of software allows users to try the program free of charge, and pay for a licence for continued use if the program is accepted. This allows software to be distributed at very low cost, with the author charging a modest licence fee which nevertheless is more than he/she would obtain if the program were being distributed commercially. Shareware in the UK has not progressed so well, with authors receiving very few licence agreements, and there has been a proliferation of fringe dealers in shareware sending out some old and, in some cases, very dubious material (such as pirated commercial disks). The principle, nevertheless, is a very sound one, and the chance to try out a program before paying for it is a very valuable privilege which might be lost if authors become discouraged.

The address for PDSL is:

> Winscombe House
> Beacon Road
> Crowborough
> E. Sussex, TN6 1UL
> Tel: (0892) 663298
> Fax: (0892) 667473

At the time of writing, membership subscriptions were £21 per annum for private membership, £69 per annum for corporate membership, and disks were copied for prices of £3.75 each (members) or £5.00 (non-members) with discounts for quantities. There is a surcharge, currently 90p for 3.5-inch 720K discs.

Other useful sources

Matmos Ltd are stockists of complete machines at excellent prices, and a wide range of add-on cards and peripherals. Contact Graham Duncan at the following address:

> Matmos Ltd
> Unit 11
> Lindfield Enterprise Park
> Lewes Road, Lindfield
> W. Sussex, RH16 2BX
> Tel: (0444) 482091 or 483830
> Fax: (0444) 484258

MicroSurgeons (Isenstein) Ltd specialise in complete systems, also at very attractive prices. Contact Colin Swift at:

> MicroSurgeons Ltd
> Deeside Industrial Estate
> Welsh Road
> Deeside
> Clwyd, CH5 2LR
> Tel: (0244) 281025
> Fax: (0244) 281161

Chapter 10

Hardware Components and Practical Work

Hardware

The hardware of electronics, meaning the chasses, covers, cabinets, switches and other external features is, for the manufacturer of consumer electronics, almost as important as the circuitry within, because the external appearance is all that the casual user can use to judge the system. For military contract work, adhering to specifications is what counts, but on any score hardware cannot be neglected, though it very often is. The task is made much easier when there is a company policy of using some particular form of hardware such as standardised board and cabinet sizes. The most difficult decision is on how to package some piece of equipment that is, for the moment, a one-off product, particularly if there is any chance that other items will follow.

The prospective user of a piece of electronics equipment first makes contact with the design when he/she tries to connect it to other equipment and to whatever power supply is used. Mains-operated equipment for domestic or office use will usually have a connected and well-tethered mains cable of the correct rating, preferably with a correctly fused three-pin plug if it is intended for the UK market. The option is to use a BS/IEC-approved three-pin plug on the chassis, with a lead that has a matching socket at one end and a suitable mains plug at the other. Mains cables can be obtained as a standard item with the BS/IEC fitting at one end and various UK or other plugs at the other. If additional equipment has to be driven, BS/IEC sockets can be used to allow power to be taken from the main unit rather than from a set of additional mains cables. Only low-power and double-insulated equipment should ever use the two-pin form of 'cassette-recorder' leads, and only if this is enforced by considerations of space or expense.

Industrial equipment will use one of the approved industrial connectors, almost certainly to the BS 4343 specification for equipment to be used in the UK. For domestic electronics, mains connections are about as standardised as anything in the use of terminals ever attains, but there is a bewildering variety of styles available for low-voltage connectors and for signal connectors. The primary aim in choosing connections should be to achieve some coordination with the equipment that will be used along with yours, and this means that you need to have, from the start of the design stage, a good idea of what amounts to *de facto* standards. If no such standards exist, try to resist making new ones because there are far too many already.

Take, for example, the low-voltage supply connector which is used for portable stereo players and for some calculators. Even in this very restricted range of applications there are two sizes of plug/socket, the 2.1 mm and the 2.5 mm, with some manufacturers using the centre pin as earth and others using the centre pin as the supply voltage pin (usually 6 V) so that there can be four variations on this design alone. Power supplies for such equipment usually cope by using a lead that is terminated in a four-way jack plug and which at the supply end is fitted with a reversible plug, often with no polarity markings. Getting the correct polarity is very often more a matter of luck than good instructions, so if this type of connector is used there should be a clear indication of polarity and also some protection against use of the wrong polarity (a diode in series) if the voltage drop can be tolerated.

The main confusion, however, exists among signal connectors, and the only possible advice here is to try to stay with industry standards for comparable equipment. For educational equipment, for example, the 4 mm plug and socket is almost universal, and for connecting UHF signals to a domestic TV receiver the standard coaxial plug and socket type should be used. Connections to TV receivers that are being used as monitors often use the SCART (Standard Connector for Audio, Radio and Television) form of Euroconnector, but for other connections, particularly for computer monitors, standards can vary widely. Fortunately, the almost universal adoption of the IBM PC standards in computing everywhere except in education (where they are most needed) ensures reasonable uniformity.

The largest range of connectors is found in the RF and video ranges, with audio coming a close second. RF connectors are used for radio transmitters, including CB and car telephone uses, and a variety of VHF and UHF work. Although these are virtually all coaxial in design they offer a wide range of fittings, whether bayonet or screw retained. The wide range of connectors reflects the wide range of VHF and UHF cables, so that you cannot necessarily use any type of connector with any type of cable. Generally, connectors are made to work with a limited range of cables, though in some cases the range can be extended by using adapters.

The range of RF connectors is intended to match the range of RF cables, of which *Figure 10.1* shows a summary only, in which some of the better-known cable types are arranged into groups. RF cables are identified by Uniradio numbers or by RG numbers. Both systems of classification are in widespread use in the UK, but the RG set, of US origin, is better known world-wide. The main measurable features of an RF cable are the characteristic impedance to which the cable must be matched in order to minimise reflections, the capacitance per metre length, and the attenuation in decibels per 10 m length at various RF frequencies. All of the cable groups illustrated are coaxial and the characteristic impedance is virtually always either 50 Ω or 75 Ω, but the attenuation characteristics differ considerably from one cable type to another. Most of the RF cables are rated to withstand high voltages between inner and outer conductors, often exceeding 20 kV.

The various connectors are likely to be used for other than RF cable connections, of course, and there is a wide range of other coaxial cables which will fit one connector type or another and to which the connectors will match well. These applications include audio, video and digital network signal applications.

The BNC range of connectors covers both 50 Ω and 70 Ω types which are manufactured for an assortment of cable sizes. All feature a bayonet locking system and a maximum diameter of about 15 mm, and both solder and crimp fittings are available. The standard range of BNC connectors can be used in the frequency range up to 4 GHz (absolute maximum

The most commonly-used types of RF cables can be classed into six groups as follows:

Gp.	Example	Impedance	pF/metre	Attenuation/frequency
A	RG58C/U	50	100	2/10 3.1/200 7.6/1000
B	RG174A/U	50	100	1.1/10 4.2/200 6.0/400
C	RG223/U	50	96	0.39/10 1.58/100 5.41/1000
D	URM70	75	67	1.5/100 5.2/1000
E	RG179B/U	75	64	1.9/10 3.2/100 8.2/1000
F	RG59B/U	75	68	1.3/100 1.9/200 4.6/1000

The UR series of cables are to UK Uniradio standards, the RG set are to US standards. Capacitance is quoted in picofarads per metre length, and the attenuation is shown in dB per 10 m metre length of cable at selected frequencies (in MHz). For example, 5.2/1000 means 5.2 db attenuation at 1000 MHZ for a 10 m length of cable.

Figure 10.1. A selection from the RG standard designations for RF cables

10 GHz) and with signal voltage levels up to 500 V peak. Terminations and attenuators in the same series are also obtainable, and there is also a miniature BNC type, 10 mm diameter, with the same RF ratings, and a screw-retained version, the TNC couplers. The connectors of this family offer a substantially constant impedance when used with the recommended cables.

The SMA series of connectors are to BS 9210 N0006 and MIL-C-39012 specifications, and are screw retained. This provides more rigidity and improves performance under conditions such as vibration or impact. The voltage rating is up to 450 V peak, and frequencies up to 12.4 GHz on flexible cable and up to 18 GHz on semi-rigid cable are usable. The VSWR (Voltage Standing Wave Ratio, ideally equal to 1.00) which measures reflection in the coupling is low at the lower frequencies but increases linearly with frequency. A typical quoted formula for semi-rigid cable coupling is $1.07 + 0.008f$ with f in gigahertz, so that for a 10 GHz signal the VSWR would be $1.07 + 0.008 \times 10 = 1.15$. The body material is stainless steel, gold plated, with brass or beryllium–copper contacts, and a PTFE insulator. Operating temperature range is $-55\,°C$ to $+155\,°C$.

The SMB (Sub-miniature Bayonet) range is to BS 9210 and MIL-C-93012B specifications and has a 6 mm typical diameter, rated for 500 V signal peak in the frequency range up to 4 GHz. The VSWR is typically around 1.4 for a straight connector and 1.7 for an elbowed type, and both solder and crimp fittings are available. SMC (Sub-miniature Screw) connectors are also available to the same BS and MIL specifications.

The older UHF plug series are also to MIL specifications, but with the limited frequency range up to about 500 MHz. These are larger connectors, typically 19 mm diameter for a plug, with screw clamping, and they are particularly well suited to the larger cable sizes. They are often also used for video signal coupling.

There are adapters available for every possible combination of RF connector, so that total incompatibility of leads should never arise. This is not a perfect solution, however, because the use of an adapter invariably increases the VSWR figure, and it is always better to try to ensure that the correct matching plug/socket is used in the first place.

Video connectors

Video connections can make use of the VHF type of RF connectors, or more specialised types. These fall into two classes, the professional video connectors intended for use with TV studio equipment, and the domestic type of connector as used on video recorders to enable dubbing from one recorder to another or from a recorder to a monitor so that the replayed picture can be of better quality than is obtainable using the usual RF modulator connection to the aerial socket of a TV receiver. Many video recorders nowadays use nothing more elaborate than an audio-style coaxial phono connector for their video as well as for their audio output.

For studio use, video connectors for a camera may have to carry a complete set of signals, including separate synchronising signals, audio telephone signals for a camera-operator, and power cables as well as the usual video-out and audio-out signals. Multiway rectangular connectors can be used for such purposes, with eight-way connections for small installations and 20-way connections for editing consoles and similar equipment. These connectors feature very low contact resistance, typically 5 m Ω. For smaller equipment, circular cross-section connectors of about 17 mm overall diameter are used, carrying ten connectors with a typical contact resistance of 14 m Ω and rated at 350 V AC.

Audio connectors start with the remarkably long-lived jacks which were originally inherited (in the old 0.25-inch size) from telephone equipment. Jacks of this size are still manufactured, both in mono (two-pole) and stereo (three-pole) forms, and either chassis mounted or with line sockets. Their use is now confined to professional audio equipment, mainly in the older range, because there are more modern forms of connectors available which have a larger contact area in comparison with their overall size. Smaller versions of the jack connector are still used to a considerable extent, however, particularly in the stereo form. The 3.5 mm size was the original miniature jack, and is still used on some domestic equipment, but the 2.5 mm size has become more common for mono use in particular.

One of the most common forms of connector for domestic audio is still the phono connector whose name indicates its US origins. Phono connectors are single channel only, but are well screened and offer low-resistance connections along with sturdy construction. The drawback is the number of fittings needed for a two-way stereo connection such as would be used on a stereo recorder, and for such purposes DIN plugs are more often used, particularly in European equipment. Many users prefer the phono type of plug on the grounds of lower contact resistance and more secure connections.

The European DIN (Deutsches Industrie Normallschaft, the German standardising body) connectors use a common shell size for a large range of connections from the loudspeaker two-pole type to the eight-way variety. Though the shell is common to all, the layouts, *Figure 10.2*, are not. The original types are the three-way and the 180° line-way connectors, which had the merit of allowing a three-way plug to be inserted into a five-way socket. Later types, however, have used 240° pin configurations with five, six and seven connectors, four-way and earth types with the pins in square format, and a five-way domino type with a central pin, along with the eight-way type which is configured like the seven-way 240° type with a central pin added. This has detracted from the original simplicity of the scheme, which was intended to make the

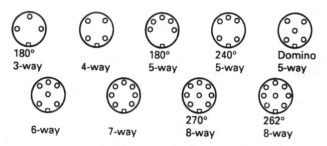

Figure 10.2. The format of the DIN connector family

connections to and from stereo domestic audio equipment easier. The more crowded layouts of plugs and sockets are notoriously difficult to solder unless they have been mechanically well designed, using splayed connectors on the chassis-mounted sockets and to some extent also on the line-mounted plugs. Use of the five-way 240° type is recommended for audio equipment other than professional-grade equipment, but only where signal strengths are adequate and risk of hum pickup is minimal. Latched connectors can be obtained to avoid the possibility of pulling the connectors apart accidentally. For low-level use, phono plugs are preferable.

For professional audio (or high-quality domestic audio) equipment, the XLR series of connectors provides multiple connections with much superior mechanical quality. These are available as three-, four-, or five-pole types and they feature anchored pins and no loose springs or set screws. The contacts are rated to 15 A for the three-pole design (lower for the others) and they can be used for a maximum working voltage of 120 V. Contact resistance is low, and the connectors are latched to avoid accidental disconnection. There is a corresponding range of loudspeaker connectors to the same high specifications. A variety of other connectors also exists, such as the EPX series of heavy-duty connectors and the MUSA coaxial connectors. These are more specialised, and would be used only on equipment that is intended to match other items using these connectors.

Computer and other digital signal connections have, at least, reached some measure of standardisation on the PC type of machines (IBM clones and compatibles) after a period of chaos. The use of edge connectors should be confined to internal connections because edge connectors are much too fragile for external use.

The Centronics connector is used mainly for connecting a computer to a printer, and it consists of a 36-contact connector which uses flat contact faces. At one time, both computer and printer would have used identical fittings, but it is now more common for the 36-pin Centronics socket to be used only on the printer. At the computer a 25-pin sub-miniature D-connection is used, usually with the socket chassis mounted. In a normal connection from computer to printer, only 18 of the pins are used for signals (including ground). The shape of the body shell makes the connector irreversible.

The same 25-pin D-connector can be used for serial connections, but more modern machines are nine-pin sub-miniature D-sockets for this purpose, since no more than nine-pin connections are ever needed. For other connections, such as to keyboards, mice and monitors, DIN-style connectors are often used, though the sub-miniature D-type connectors are also common. The D-type connectors are widely available in a range of sizes and with a large range of accessories in the form of casings, adapters and tools, so that their use for all forms of digital signals is strongly recommended.

There are now standard DIN fittings for edge connectors, including the more satisfactory indirect edge connectors that have now superseded the older direct style. The indirect connectors are mounted on the board and soldered to the PCB leads, avoiding making rubbing contacts with the board itself.

Control knobs and switches

There is as great a variety in control knobs and switches as in terminals and connectors. Control knobs for rotary potentiometers are available in a bewildering range of sizes and styles, mostly using grub-screw fastening, though a few feature push-on fitting. For all but the least costly equipment, a secure fastening is desirable, but the traditional grub-screw is not entirely satisfactory because it can work loose and can cause considerable delay when knobs have to be removed for servicing work. A more modern development is collet-fitting, using a split collet over the potentiometer shaft which is tightened down by a nut. The recess for the collet nut is then covered by a cap which can be colour coded or moulded with an arrow pointer. This is a much more satisfactory form of fitting.

A more specialised form of knob allows multi-turn use, so that 10 to 15 turns of the dial will be needed to rotate the potentiometer shaft from one end stop to the other. These multi-turn dials can use digital or analogue readouts and are normally located by a grub-screw with an Allen (hexagon) head.

Switches

Switches are required to make a low-resistance connection in the ON setting, and a very high-resistance insulation in the OFF setting. The resistance of the switch circuit when the switch is on (made) is determined by the switch contacts, the moving metal parts in each part of the circuit which will touch when the switch is on. The amount of the contact resistance depends on the area of contact, the contact material, the amount of force that presses the contacts together, and also in the way that this force has been applied.

If the contacts are scraped against each other in a wiping action as they are forced together, then the contact resistance can often be much lower than can be achieved when the same force is used simply to push the contacts straight together. In general, large contact areas are used only for high-current operation and the contact areas for low-current switches as used for electronics circuits will be small. The actual area of electrical connection will not be the same as the physical area of the contacts, because it is generally not possible to construct contacts that are precisely flat or with surfaces that are perfectly parallel when the contacts come together.

A switch contact can be made entirely from one material, or it can use electroplating to deposit a more suitable contact material. By using electroplating, the bulk of the contact can be made from any material that is mechanically suitable, and the plated coating will provide the material whose resistivity and chemical action is more suitable. In addition, plating makes it possible to use materials such as gold and platinum which would make the switch impossibly expensive if used as the bulk material for the contacts. It is normal, then, to find that contacts for switches are constructed from steel or from nickel alloys, with a coating of material that will supply the necessary electrical and chemical properties for the contact area.

Switch ratings are always quoted separately for AC and for DC, with the AC rating often allowing higher current and voltage limits, particularly for inductive circuits. When DC through an inductor is decreased, a reverse voltage is induced across the inductor, and the size of this voltage is equal to inductance multiplied by rate of change of current. The effect of breaking the inductive circuit is a pulse of voltage, and the peak of the pulse can be very large, so that arcing is almost certain when an inductive circuit is broken unless some form of arc suppression is used.

Arcing is one of the most serious of the effects that reduce the life of a switch. During the time of an arc very high temperatures can be reached both in the air and on the metal of the contacts, causing the metal of the contacts to vaporise and be carried from one contact to the other. This effect is very much more serious when the contacts carry DC, because the metal vapour will also be ionised, and the charged particles will always be carried in one direction. Arcing is almost imperceptible if the circuits that are being switched run at low voltage and contain no inductors, because a comparatively high voltage is needed to start an arc. For this reason, then, arcing is not a significant problem for switches that control low voltage, such as the 5 V or 9 V DC that is used as a supply for solid-state circuitry, with no appreciable inductance in the circuit. Even low-voltage circuits, however, will present arcing problems if they contain inductive components, and these include relays and electric motors as well as chokes. Circuits in which voltages above about 50 V are switched, and particularly if inductive components are present, are the most susceptible to arcing problems, and some consideration should be given to selecting suitably rated switches, and to arc suppression, if appropriate.

The normal temperature range for switches is typically $-20\,°C$ to $+80\,°C$, with some rated at $-50\,°C$ to $+100\,°C$. This range is greater than is allowed for most other electronic components, and reflects the fact that switches usually have to withstand considerably harsher environmental conditions than other components. The effect of very low temperatures is due to the effect on the materials of the switch. If the mechanical action of a switch requires any form of lubricant, then that lubricant is likely to freeze at very low temperatures. Since lubrication is not usually an essential part of switch maintenance, the effect of low temperature is more likely to alter the physical form of materials such as low-friction plastics and even contact metals.

Flameproof switches must be specified wherever flammable gas can exist in the environment, such as in mines, in chemical stores, and in processing plants that make use of flammable solvents. Such switches are sealed in such a way that sparking at the contacts can have no effect on the atmosphere outside the switch. This makes the preferred type of mechanism the push-on, push-off type, since the push-button can have a small movement and can be completely encased along with the rest of the switch.

Switch connections can be made by soldering, welding, crimping or by various connectors or other plug-in fittings. The use of soldering is now comparatively rare, because unless the switch is mounted on a PCB which can be dip-soldered, this will require manual assembly at this point. Welded connections are used where robot welders are employed for other connection work, or where military assembly standards insist on the greater reliability of welding. By far the most common connection method for panel switches, as distinct from PCB-mounted switches, is crimping, because this is very much better adapted for production use. Where printed circuit boards are prepared with leads for fitting into various housings, the leads will often be fitted with bullet or blade crimped-on connectors so that switch connections can be made.

Cabinets and cases

The variety of cabinets and cases is as wide as that of the other hardware components. Small battery-operated equipment can be housed in plastic cases, particularly one-off or developmental circuits, but for the production of equipment for professional use, some form of standard casing will have to be used. As so often happens, industry standards have to be obeyed.

The old 19-inch rack standard for industrial equipment has now

U-number	Height	Width	Depth
40U	1920	647	807
34U	1620	600	600
27U	1290	600	600
20U	998	600	639
12U	619	600	639
6U	230	88*	160
3U	95	88*	160

(typical dimensions in mm)
*Smaller cases can be specified as 10E width (38 mm) or 20E width (88 mm)

Figure 10.3. The standard set of cabinet dimensions (U-set)

become the IEC 297 standard, with cabinet heights designated as **U** numbers – the corresponding millimetre measurements are shown in Figure 10.3. These cabinets can be supplied with panels, doors, mains interlocks, top and bottom panels, fan plates, and supports for chasses, providing ample space for internal wiring and cooling. Internal chasses in the form of racks and modules can be fitted, usually in 3U and 6U sizes.

Smaller units are accommodated in **instrument** cases, of which the range is much larger. There is a range of cases which will fit the 19-inch units from the standard rack systems so that identical chassis layouts can be used either in racks or in the smaller cases. The more general range of casings cover all sizes from a single card upwards, and also down to pocket calculator sizes. Cases can be obtained with carrying handles for enclosing portable instruments, or for bench or desk use. Many casings can be obtained in tough ABS plastics or in die-cast metal form. Some further degree of standardisation is emerging, as far as European equipment is concerned, as a DIN standard 43700 for small cases and boxes along with plug-in modules that fit inside.

Packages for semiconductors

Passive components such as resistors and capacitors require only two terminals, and there is no need to identify which is which. Transformers will need at least three, usually four, terminals, and these are usually identified by markings on the transformer itself or on paperwork that accompanies the transformer, or on service sheets for the equipment in which the transformer is wired. The main problems of packaging therefore apply to semiconductors, and this has been made particularly complicated by the very large numbers of different connecting arrangements that have been devised in the past. By comparison, digital i.c.s have almost reached a reasonably standardised situation. The reason for packaging is that semiconductors are very small, and the chip itself, whether of a diode, transistor or i.c., would be almost impossible to work with and very easily damaged. The chips are therefore mounted into metal or plastic, and thin wire connectors welded to the semiconductor terminals and to the wires or studs that will form the terminals of the package. The chip itself is usually surrounded by some inert material such as silicone jelly or plastic, to prevent contamination from the air. In addition, some means of carrying heat from the semiconductor may need to be used, particularly for power-dissipating devices.

Starting with the simplest types of semiconductors, small diodes are invarably marked at the cathode terminal. The marking may be a red dot or ring, but whatever type is used, the fact that it is on the cathode is sufficient for identification. One of the main problems about diodes is marking the type letters or number. The very small size of a diode can make these type letterings very difficult to read, particularly when the diode is connected into a circuit, so that a circuit diagram labelled with semiconductor types is particularly useful. Larger diodes for rectification often make use of the stud type of packaging, with the cathode connected to a mounting stud (*Figure 10.4*). Another common packaging is of four diodes in a bridge connection with the four terminals provided for a full-wave connection (*Figure 10.5*). When this packaging is used, there may be a bolt-hole provided so that the bridge can be bolted to a metal fin for heat dissipation.

The packages for transistors can be divided into small-signal types and power types, and similar packages are used for thyristors. The main small-signal packages encountered nowadays are the TO72, TO18, TO5, TO39, TO92, E-line and Silect, and *Figure 10.6* indicates the variations that these represent. A considerable complication about such packages as the TO72 is that two transistors using the same packaging may not have identical connections. The only way of finding out the connections is to know the transistor type and look up the connections in a hand-book, though for a transistor connected in a circuit, the components connected to the terminals will often provide useful clues as to what the connections

Figure 10.4. Typical stud-mounting diodes use large-current rectifiers

Figure 10.5. A set of four diodes packaged as a bridge circuit

are. Determining connections can be particularly difficult when transistors are unmarked. A lot of equipment uses transistors bought in bulk from the manufacturers and marked only with factory internal codes. Unless there are manuals for the equipment that name equivalent transistor types and/or show connections, the servicing of such equipment can be very difficult.

The main principle in packaging is to provide for recognition of leads, and to make the transistor easy to connect into circuit, particularly by automatic means. The case of the transistor is usually made asymmetrical in some way, by a tag, or by flattening one side, so that it should be possible to insert the transistor only one way round in automatic insertion machines. Case shapes like the E-line are not always so easy to check, because the difference between the flat side and the rounded side is not easy to see, particularly when the transistor is one of a large number packed on a printed circuit board.

Commonly used packaging styles for power transistors are illustrated in *Figure 10.7*. The important feature of any packaging for these devices is heat dissipation, and so most of the packages feature metal tabs or studs which are in good thermal contact with the collector, drain or anode of the device. Here again, some types of package come in several connection

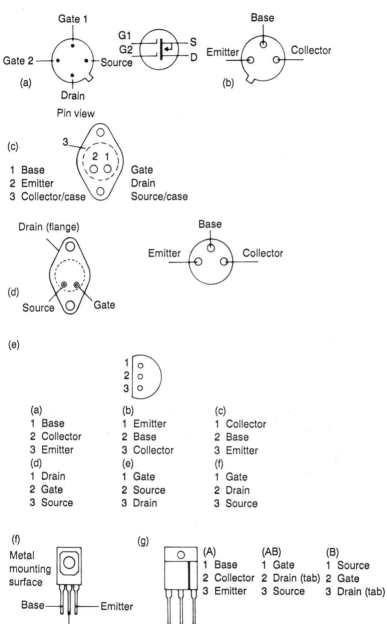

Figure 10.6. Some of the more common transistor packages

(h) Pin view

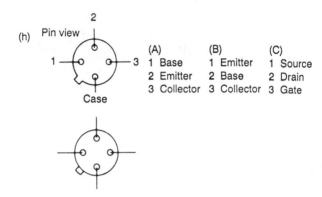

	(A)	(B)	(C)
1	Base	Emitter	Source
2	Emitter	Base	Drain
3	Collector	Collector	Gate

(i)

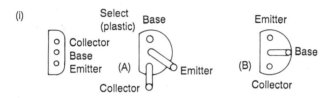

Figure 10.6 (cont.)

Figure 10.7. Some packaging styles for power transistors and thyristors

styles. Both thyristors and Triacs can also use the type of stud package that is used for some rectifier diodes, but with two terminals in addition to the stud connection, *Figure 10.8*.

Integrated circuit packages

The types of packages used for i.c.s can show an even greater degree of variation. Some early i.c.s were packaged like transistors, but with up to nine leads coming out from the can. This type of packaging is still used for some high-frequency amplifier i.c.s. Later types have tended to use the block style of packaging, with the chip contained in a flat rectangular slab of plastic with the leads on each side of the slab. The most familiar package of this type is the dual-in-line (DIL) package, illustrated in *Figure 10.9* in its eight-pin form. The pin spacing for this type of pack is standardised so that automatic machinery can be used for drilling printed circuit boards and so that automatic inserting machines can be used to populate the boards. The DIL principle has been bent slightly for some large chips, such as microprocessors, to provide for 40-pin chips with a 0.6-inch separation between lines. Some more recent microprocessor types have used longer types of DIL packaging in order to accommodate up to 64 pins. A few recent microprocessors have appeared with a square slab that uses a double-decker row of pins along each of the four edges.

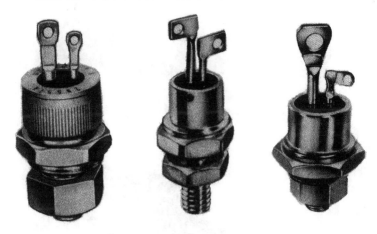

Figure 10.8. The stud form of packaging used for thyristors and Triacs

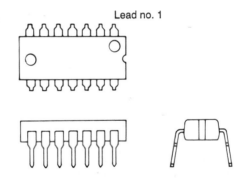

Figure 10.9. The form of the DIL package used for both linear and digital i.c.s

The names of pin-grid arrays or leadless chip-carriers have been applied to two styles of package that are already in use, *Figure 10.10*.

The main aberrations in packaging are found in linear i.c.s, particularly where power dissipation is involved. A power amplifier i.c. must provide for several pin connections and also for heat dissipation, and several types incorporate heat sinks as part of the structure. Integrated circuits of this type are very difficult to replace if the original type is no longer available, because although the functions of the chip may be easy to replace, the physical shape and mounting is not, and adapter boards may be needed. Many of the common types of linear i.c.s, such as OPamps, are packaged in DIL form, often with eight pins.

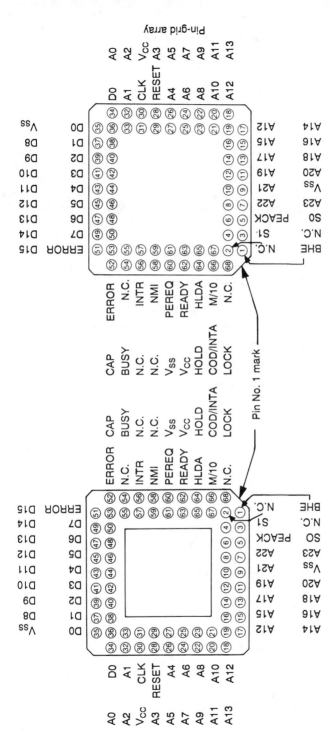

Figure 10.10. Pin-grid and leadless chip-carrier packages for large-scale i.c.s such as microprocessors and allied chips

Handling

There is an important distinction between transistors and i.c.s in the sense that transistors are always soldered into circuits, but i.c.s can be either soldered to a printed circuit board or plugged into holders that are soldered to the board. The plug-in construction for i.c.s has been used extensively for small computers, particularly when the design makes it possible to expand the memory of the machine by plugging in more memory i.c.s. The use of i.c. holders is not necessarily helpful either from the reliability point of view or that of servicing, however. From the production point of view, the use of i.c. holders means that another step is needed – insertion of i.c.s – before the board is ready for use. The use of insertion tools in production makes the operation relatively easy, but for small-scale runs, insertion of i.c.s without inserting tools can easily cause pin damage. The most common form of damage is for one pin which has not been correctly lined up to hit the side of the holder and be bent under the body of the i.c. as it is inserted. Visual inspection often fails to spot this, and it is only when pin voltages are read using a logic probe that the disconnection will be noticed. This mishandling is even easier to do when an i.c. is being replaced, and several manufacturers take the view that it is more satisfactory to solder-in all i.c.s, since the reliability of i.c.s is such that replacement is unlikely to be needed.

There is a lot to be said for this, because holders are by no means completely satisfactory, and as much time can be spent in removing and inserting i.c.s from and into holders as would be spent in snipping the pins, removing the remains and soldering in a new i.c. A further point is that inserting an MOS i.c. into a holder is more likely to cause electrostatic damage than soldering the i.c. If all of the pins are not placed in contact with the holder at the same time, there is a risk that one pin which is isolated could be touched. Though the built-in diodes of MOS i.c.s will generally protect against damage, the risk is higher than that of soldering. It is possible to solder-in an i.c. with a wire wrapped around all the pins high up on the shanks. This shorts all of the pins together, preventing any risk of electrostatic damage while the i.c. pins are being soldered. This is particularly useful when an i.c. is being replaced during servicing, because the pins will generally be soldered one at a time in such a case. The shorting wire can be removed after all the soldered joints have been checked.

The tendency now on computers is to use surface mounting wherever possible, and the minimum of socketing. On modern PC machines, a socket will be provided for a co-processor, usually a floating-point maths unit. The memory of the machine will be housed in strips designated as SIMM (Single Inline Memory Module) or SIPP (Single Inline Pin Podule) which use edge or pin connections respectively. The main board (or motherboard) of the computer will contain holders for SIMM or SIPP units, and the memory units themselves use surface-mounted or conventional memory chips that are soldered in place. This allows the memory to be inserted and changed by plugging in single units rather than a large number of chips. At one time, adding 1 M to the memory of a PC computer could involve plugging in 36 chips (256×1-bit chips, using 8 bits for each memory byte and 1 parity bit for checking the integrity of the other 8 bits). It is now much more common for this to be done by plugging in one SIMM or SIPP unit, and 4 Mb SIMM/SIPP units are also available.

Desoldering of an i.c. is needed only if there is some doubt about a fault. If the chip is known to be faulty, then desoldering is a waste of time – it is always easier and less harmful to other i.c.s to snip the pins of the defective i.c. and then remove them one by one from the board with a hot soldering iron and a pair of pliers. On the few occasions when a chip has to be desoldered and kept in a working state, the use of some sort of desoldering tool is helpful. The type of extended soldering iron bit that covers all the pins of an i.c. is by far the most satisfactory method, but a separate head is needed for each different size of i.c. This technique is particularly useful if the chip is to be tested in a separate tester or in another circuit. It is seldom satisfactory to unsolder an i.c. and expect to test it in a circuit that makes use of holders, because the presence of even only a film of solder on the pins of an i.c. makes it very difficult to insert into a holder. If the need to remove an i.c. by desoldering is a task that is seldom needed, then when it does arise it can be tackled by using desoldering braid. This is copper braid which is laid against a soldered joint. When a hot soldering bit is held against the braid, the solder of the joint will melt and will be absorbed by the braid. The braid can be removed, leaving the joint free of solder. The piece of braid that is now full of solder can then be cut off, and another piece of braid used. This is a slow business, since it has to be repeated for each pin of the i.c., but the printed circuit board is left clean and in good condition. The snag is that a hot iron is being applied to the board many times, and this can cause overheating of other components. See later in this chapter for working on SMT components.

Heat dissipation

Heat dissipation from i.c.s and transistors is a critical feature of many circuits, and failure to dissipate heat correctly can be the cause of many failures. It is not generally appreciated that the memory boards of computers can run hot, and that by placing one board over another it is possible to reduce the cooling to such an extent as to cause failure. Many of the popular small computers can be fitted with add-on boards that look as if they had been designed by a failed plumber, and it is a tribute to the sturdiness of modern i.c.s that more faults do not occur. Even the original designs can show evidence that the designers did not spend much time wondering where the power would be dissipated. There is little that can be done if the original design is prone to overheating, other than cut a slit in the box and attach a cooling fan. Having said that, I should point out that I have tested most of the present generation of small computers for up to 10 hours per day continuous running and have never encountered problems with the basic design, as distinct from add-ons, even on machines with no cooling fan.

The main heat dissipation problems arise when power transistors of i.c.s are in use, and are attached to a heat sink. Once again, well-designed original equipment gives very little trouble unless there has been careless assembly of transistors or i.c.s on to heat sinks. This can also be a cause of a newly repaired circuit failing again in a very short time. The problem arises because the flow of heat from the collector of a transistor to the metal of a heat sink is very similar to the flow of current through a circuit. At any point where there is a high resistance to the flow of heat, there will be a 'thermal potential difference' in the form of a large temperature difference. This can mean that the heat-sink body feels pleasantly warm, but the collector junction of the transistor is approaching the danger level. The problem arises because of the connections from one piece of metal to another. Any roughness where two pieces of metal are bolted together will drastically increase the thermal resistance and cause overheating. If either the transistor/IC or the heat sink has any trace of roughness on the mounting surface, or if the surface is buckled in such a way as to reduce the area of contact, then some metalwork with a fine file and emery paper will be needed. In addition, silicone heat-sink grease should always be used on both surfaces before they are bolted together. If the grease is not pressed out from the joint when the joints are tightened, this is an indication that contact is not good. In addition, though, the heat-sink grease is quite a good heat conductor, and will greatly reduce the thermal resistance where two metal surfaces are bolted together. Failure to use heat-sink grease, either at original assembly or later when a transistor or i.c. is replaced, will almost certainly cause trouble. One of the problems about poor heat-sinking is that the problem that it causes may be seasonal, so that the equipment behaves flawlessly throughout the 364-day British winter, only to expire mysteriously on the day of summer.

Constructing circuits

The methods that are used industrially to construct circuit boards in large numbers are quite different to the methods that have to be used for experimental or one-off circuits. One factor that is common to all constructional methods, however, is that the circuit diagram is a way of showing connections which does not give any indication of how the components can be physically arranged on a board.

The main difference between circuit diagrams and layout diagrams is that on a layout diagram any crossing of connecting leads has to be avoided. A component such as a resistor or capacitor can cross a circuit-connecting track because on a conventional printed circuit board (PCB) the component will be on the opposite side of the board from the track (but see Surface mounting, later in this chapter). The simplest circuits to lay out are discrete transistor amplifiers; the most difficult are digital circuits in which each chip contains a large number of separate devices and which are generally laid out on double-sided boards so as to make the geometry of the boards possible. Double-sided boards require dummy pads, known as vias, which are used only to connect a lead on one side with a lead on the other, and if a layout is not a good one, there may even need to be connections made by wire leads.

Small-scale circuits can be laid out manually, using cardboard cut-outs of the components, with connections and internal circuits marked, on a large sheet of tracing paper or on transparent plastic which has been marked with a pattern of dots at 0.1-inch centres. The designing is done looking at the component side of the board, and will start by roughing out a practicable layout which does not require any track crossovers (but see Chapter 9 regarding computer layout programs). At this stage, it is important to show any interconnecting points, using edge connections or fixed sockets, because it must be possible to take leads to these connectors without crossovers. This can be quite difficult when the connection pattern is fixed in advance, as for example when a standard form of connection like a stereo DIN socket or a Centronics printer socket is to be used.

This layout can then be improved, with particular attention paid to durability and servicing. The positions of presets and other adjustments will have to be arranged, along with points where test voltages can be measured, so that servicing and adjustment will be comparatively easy. Signal lines may have to be re-routed to avoid having some lines running parallel for more than a few millimetres (because of stray capacitances), or to keep high-impedance connections away from power supply leads. It is normal, however, for bus lines on microprocessor circuitry to run parallel. Some tracks that carry RF may need to be screened, so that earthed tracks must be provided, possibly on each side, to which metal screens can be soldered. Components which will run hot, such as high-wattage resistors, will have to be mounted clear of other components so that they do not cause breakdown because of overheating in semiconductors or capacitors. One way of achieving this is to use long leads for these components so that they stand well clear of the board, but this is not always a feasible solution if several boards have to be mounted near to each other.

More extensive circuits can be planned by computer, see Chapter 9, so that the PCB pattern is printed out after details of each component and each join in the circuit have been typed into the computer. Even computer-produced layouts, however, may have to be manually adjusted to avoid having unwanted stray capacitances appearing between components. Either method will have to take account of the physical differences that can exist between similar components, such as length of tubular capacitors and the difference between axial and radial lead positions.

The simplest form of construction for a one-off circuit is the matrix stripboard. This can be obtained in a wide range of sizes, up to 119×455 mm (4.7×18 inches approx.). The traditional type of stripboard in both 0.15-inch and 0.1-inch pitch are still available. These are always single sided, and are suited mainly to small-scale analogue circuitry. For digital circuits there is a range of Eurocard prototyping boards, either single or double sided. For connections between strips on opposite sides, copper pins are available which can be soldered to each track, avoiding the difficulties of making soldered-through connections on such boards. Boards can also be obtained in patterns such as the IBM PC expansion card or the Apple expansion card forms. Tracks can be cut by a 'spot-cutter' tool which can be used in a hand or electric drill, thus allowing components like DIL i.c.s to be mounted without shorting the pin connections, *Figure 10.11*. Once these cuts have been made, the components can be soldered on to the board and the circuit tested. Arrangement of components for really small circuits can be tested in advance by using a **solderless breadboard**, which allows components to be inserted and held by spring clips, using a layout which is essentially the same as for matrix stripboards.

The circuit needs to be marked out in nodes, see Chapter 9, and a separate strip of connector assigned to each node. For complex circuits, this can be easier if the node numbers are also marked on the strips, using typists' erasing liquid such as Tippex to make a white surface on which pencil marks can be made. The components can then have their leads bent and cut to fit, and can be soldered into place. Components such as DIL i.c.s can be mounted without any need to trim leads, but remember to cut the strips that would otherwise short out pins.

Soldering on to stripboard is usually easy, but some components may need some heat-shunting. Avoid old pieces of stripboard whose copper surfaces have become tarnished, because good soldering on to such surfaces is not easy, and is likely to need a hot iron with risk to semiconductor components. Always use clean board, tinned if available. The layout that is used on these boards can be used also as a pattern for manufacturing PCBs for mass production.

For larger-scale work, PCBs must be produced which will later be put into mass production. A board material must be used which will be strong and heat resistant, with good electrical insulation. The choice will usually be between plastic-impregnated glass fibre board, or SRBP (Synthetic Resin-bonded Paper), though some special-purpose circuits may have to be laid out on ceramic (like porcelain) or vitreous (like glass) materials in

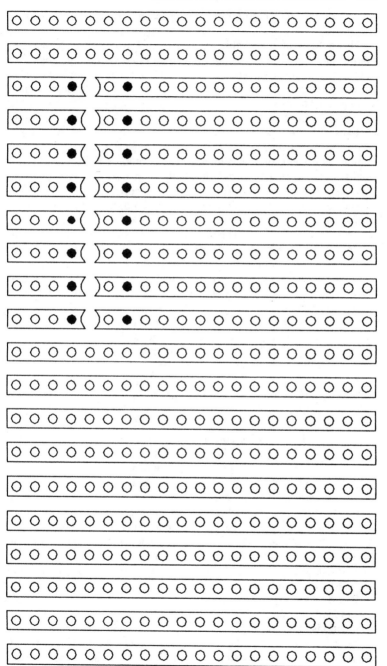

Figure 10.11. Tracks cut on a stripboard to prevent shorting i.c. pins

order to cope with high-temperature use and flameproofing requirements.

The pattern of the circuit tracks has to be drawn on a piece of copper-laminated board, using a felt-tipped pen which contains etch-resistant ink. For experimental uses, circuit tracks can be applied directly by using etch-resistant transfers. Standard patterns include lines of various thicknesses, i.c. and transistor pads, and pads for mounting connectors, together with a variety of curves, dots, triangles and other patterns. Another option is to work from a transparency of the pattern, using light-sensitive etch-resistive material which is then 'developed' in a sodium hydroxide solution. When the ink or other etch-resist is dry, the board is etched in a ferric chloride bath (acid hazard – wear goggles, gloves and an apron) until all of the unwanted copper has been removed. The layout of a circuit on to copper laminate board should start by making a drawing on tracing paper or transparent film. Components, or cardboard cut-outs can then be placed on the drawing to show sizes, and to mark in the mounting pads to which the leads will be soldered. This is done with the drawing representing the component side of the board, but the tracks can now be drawn in as they will exist on the copper side – this means that the actual appearance of the component side will be the mirror-image of your layout.

The drawing will then have to be transferred to the copper. This can be done manually, using an etch-resistant ink as described earlier, or by photographic methods. The board can then be etched, thoroughly cleaned, drilled and then the components mounted and soldered into place. For the hand-made etched board, all traces of the resist material and the etching solution have to be removed by washing and scrubbing with wire wool.

The board can then be drilled, using a 1-mm drill, and the components inserted. The final action is soldering, using an iron with a small tip. The boards are heat resistant, not heatproof, so that soldering should be done fairly quickly, never keeping the iron in contact with the copper for too long. Excessive heating will loosen the copper from the plastic board, or burn the board, and if the copper has been cleaned correctly and all component leads are equally clean, soldering should be very rapid. Chemical tinning solutions can be used to treat the copper of the board so that soldering can be even more rapid. Remember that excessive heat will not only damage the board but also the more susceptible components like semiconductors and capacitors. The time needed to obtain a good soldered joint should not exceed a few seconds.

Boards for larger-scale production are undrilled and completely covered with copper, which will then be etched away into the pattern of connections that is needed. In a mass-production process, the pattern of etch-resist is placed on to the copper by a silkscreen printing process. The copper is then etched in baths which are maintained at a constant high temperature, and the boards are washed thoroughly in water, followed by demineralised (soft) water, and then finally in alcohol so as to make drying more rapid. The holes are then drilled for the component leads. For mass production, all of these processes are completely automated, and the assembly of the components on to the board and subsequent soldering will also be totally automatic.

Circuit boards for commercial use have been of almost a stereotyped pattern until the comparatively recent rise in the use of surface-mounted components. The standard board backing materials are either SRBP (Synthetic Resin-bonded Paper) or glass and epoxy resin composites, with copper coating. Many suppliers offer such boards with the copper already coated with photoresist, saving considerable time and effort for small batches. These pre-coated boards must be stored carefully, preferably at low temperatures between 2 °C and 13 °C, and have a shelf life which is typically 1 year at 20 °C. The maximum allowable temperature is 29 °C.

Board sizes now follow the Eurocard standards of 100 mm × 160 mm, 100 mm × 220 mm, 233.4 mm × 160 mm and 233.4 mm × 220 mm; and there are also the older sizes of 203 mm × 95 mm (8 × 3.75 inches) and 304.8 mm × 457.2 mm (12 × 18 inches). Boards can be obtained with edge-connecting tongues already in place – these must, of course, be masked when the main board is etched. When boards are bought uncoated, photoresist can be sprayed as an aerosol for small-scale production or R&D applications.

Many commercial PCBs, particularly for computer or other digital applications, are double sided, with tracks on the component side as well as on the conventional track side. Where connections are needed between sides, plated-through holes are used. These are holes which have copper on each side and which have been electroplated with copper so that the holes have become partly filled, making a copper contact between the sides. These connections are strengthened when the board is soldered.

The use of double-sided board is particularly important for digital circuits where a single-sided board presents difficulties because of the need to cross leads. The use of a well-designed double-sided board can solve these problems, but care needs to be taken over capacitances between tracks that are on opposite sides of the board.

Surface mounting

Surface-mounting components and boards are now increasingly being used in equipment. The principle of surface mounting is not new – surface-mounting boards for amateur use were on sale in 1977, when they were demonstrated under the name of 'blob-boards' at several exhibitions. The technique, known as SMT (Surface Mounting Technology) has now spread to professional equipment and has resulted in the manufacture of a whole range of components that are designed specifically for this type of fixing. Components for surface mounting use flat tabs in place of wire leads, and because these tabs can be short the inductance of the leads is greatly reduced. The tabs are soldered directly to pads formed on to the board, so that there are always tracks on the component side of the board. Most SMT boards are two sided, so that tracks also exist on the other side of the board.

The use of SMT results in manufacturers being able to offer components that are physically smaller, but with connections that dissipate heat more readily, are mechanically stronger and have lower electrical resistance and lower self-inductance. Some components can be made so small that it is impossible to mark a value or a code number on to them. This presents no problems for automated assembly, since the packet need only be inserted into the correct hopper in the assembly machine, but considerable care needs to be taken when replacing such components, which should be kept in their packing until they are soldered into place. Machine assembly of SMT components is followed by automatic soldering processes, which nowadays usually involve the use of solder-paint (which also retains components in place until they are soldered) and heating by blowing hot nitrogen gas over the board. Solder-baths are still used, but the hot-gas method causes less mechanical disturbance and can also allow heat-sensitive components to be shielded.

Considerable care is needed for hand-soldering and unsoldering SMT components. A pair of tweezers can be used to grip the component, but it is better to use a holding arm with a miniature clamp, so that both hands can be free. The problem is that the soldering pads and the component itself can be so small that it is difficult to ensure that a component is in the correct place. Desoldering presents equal difficulties – it is difficult to ensure that the correct component is being desoldered, and almost impossible to identify the component after removal; a defective SMT component should be put into a 'rejects' bin immediately after removal.

Testing and trouble-shooting

Supplies and signals

Wherever semiconductors are used, great care must be taken over the polarity of supply voltages and signals. Reversal of supply polarity to most i.c.s, for example, will result in instant destruction, and no form of fusing can avoid this damage. The risk is greatest when several boards are joined to a supply by means of plugs and sockets. The plugs and sockets must be polarised so as to avoid accidental reversal, but even this may not be enough. If an attempt is made, for example, to plug in a supply while the supply is switched on, it is possible that a contact can be made even though the pins of the plug cannot be inserted into the socket. The semiconductors on the circuits board will therefore have been damaged. Even if the power supply has been switched off, the capacitors may be charged, and the attempt to make the incorrect connection will be enough to cause damage. When plugs and sockets are used in this way, power supplies must be switched off before disconnecting, and the power supply capacitors discharged. The power supply should never be switched on while it is disconnected from the other boards, and no attempt should be made to switch on until all plugs have been correctly fitted into their sockets.

The provision of correct signals polarities and amplitudes is equally important, because it is easy to vaporise the base junction of a transistor by connecting it to a large signal pulse. Once again, well-designed equipment will provide very different connectors for inputs and outputs so that signals that are at very different levels will use very different connectors, and some care in assembling the connectors will avoid any problems. The main risk comes during servicing actions that involve signal generators. Servicing of the passive type that makes use of meters and oscilloscopes only involves no risks of this type, other than the possibility of causing shorting because of the careless use of probes. When signal generators are used, however, you have to make sure that the polarity and the size of signals will be appropriate to the part of the circuit to which they are applied. As a precaution, signal generators should always be set to provide the minimum possible signal when they are first switched on, and the signal polarity that is needed should always be checked. The important point here is to avoid causing 'service-induced faults'. One example is connection of a signal generator to the input of an i.c., and trying to find the output signal from an output. One common mistake is to connect up, apply a very small signal and to keep increasing the signal amplitude in the hope of seeing an output. It may be that there is a fault at some point between input and output which makes it impossible to obtain an output from the input that is being used, so that increasing the input signal amplitude will simply burn out this input, creating yet another fault in the circuit. A good knowledge of the circuit is the main safeguard here. If you know that an input of 5 mV should be needed at some point, then you should apply a signal of not more than this amplitude. If no output is seen, then instead of increasing the signal amplitude, you should shift the oscilloscope probe to a point nearer the input, until a signal can be seen. This enables the true fault to be discovered before another fault is created. For digital equipment, it is much easier to avoid incorrect signal amplitude or polarity, because so many digital systems used the standard $+5\,V$ logic signal.

Testing circuits

For a large range of general-purpose circuits, particularly linear circuits, test equipment can often amount to no more than a multimeter fitted with a set of probes. You must be aware, however, of the effect of the meter on the circuit, since this will considerably affect the readings that will be obtained. The problem is that all meters other than a few specialised electrostatic types will cause a drain of current from the circuit, and the voltage that is being measured will be reduced by the amount of voltage drop caused by this current. Many analogue multimeters, for example, use 100 μA movements. If you are measuring the voltage at a point fed by a 10 kΩ resistor, the voltage drop for 100 μA is 1 V, not exactly a negligible error. This amount of current would flow only if the meter were reading full scale, so by making measurements with higher ranges, so that the meter deflection is less, this type of error is decreased. A better solution is to use a meter with a lower current requirement, and many multimeters use 50 μA or even 20 μA movements for this purpose. The current requirement of the meter is often quoted in a roundabout fashion as the figure of ohms per volt. This is the reciprocal of the maximum deflection current for the movement, so that a 100 μA movement yields an ohms-per-volt figure of 10 Ω/V.

For a meter with several ranges, the total resistance is the ohms-per-volt figure multiplied by the voltage range, so that for a meter with 10 Ω/V on its 25 V range the total resistance is $25 \times 10\,\Omega = 250\,\Omega$. The general rule for using such a meter is that its total resistance should be at least ten times greater than the resistance of the circuit it is checking. The resistance of the circuit point is taken as its equivalent resistance with all supply lines earthed, so that a circuit consisting of a 2 MΩ resistor in series with a 1 MΩ resistor between $+6$ V and earth is taken as having resistance of 667 Ω, too high for the meter in this example.

The alternative is to use a digital meter whose input impedance is fixed, usually at 10 MΩ. The higher input resistance, which is generally constant except when range-extending probes are used, reduces the effect of the meter on the circuit and allows the meter to be used in a larger range of circuits with more confidence. This should not be taken to mean that digital meters are necessarily more precise in their readings than analogue meters. The linearity of conversion from analogue to digital determines the precision, and just because a meter shows a reading with several decimal places should not be taken to imply that the voltage is measured so precisely. In addition, because digital meters sample the input voltage several times a second, the lower-order digits may flicker when the voltage being read is not perfectly steady. Remember also that a digital meter reading cannot be interpolated. On an analogue meter it is often possible to guess that a needle position is half-way between two marked positions, but on a digital meter this is not possible.

For test and service applications, DC voltage readings are much more commonly made than current readings. These readings have to be made with some caution, and the effect of the DC resistance of the meter is one point that must be considered. Another is that the effect of signal currents must be considered, because the meter will have an effect on impedance to signals. Voltage-level testing is often specified to be in conditions of zero signal, though in some circuits the voltage can be measured correctly only when the measuring point is bypassed to signal frequencies. Where voltage checking points are included in a circuit, measurements can usually be taken with no particular precautions. As always, domestic TV equipment needs the greatest number of precautions and it is always desirable (often essential) to work from a mains-isolating transformer because of the risks of working on equipment whose chassis is live to mains.

For some equipment, notably digital equipment, DC voltage readings are almost meaningless except for supply lines, but where such measurements can be made they form a useful first line of diagnosis for fault conditions. Some care is needed, however, to interpret readings. If the voltage at a test point is stated by a manufacturer to be 1.5 V, does a fault exist if this reading is 1.4 V, or if it is 1.9 V? Some datasheets show the permitted tolerances, and it is quite common to find that nearly every voltage is close to the permitted limits, so that when no tolerances are shown some common-sense has to be used. In general, the tolerance of resistor values provides some clue. If all resistors are of 10% tolerance you cannot expect voltage-level tolerances to be closer, and the only thing to look out for is that a low voltage level is not allowing a stage to be biased off. Happily, many fault conditions cause a considerable change of voltage readings.

Current readings are seldom made, and most circuits are constructed so that making current readings is difficult if not impossible. To make a current reading the circuit must be broken so that the meter can be connected across the break points, and modern PC boards cannot easily provide for this except by using jumper links. A few circuits provide for calculating current by measuring the voltage drop across a small resistor placed in the current path.

Using the CRO

Though DC voltage measurements form the first line of checking and fault diagnosis, signal waveforms may need to be traced when there are no clear indications from the voltmeter. The modern CRO used for diagnosis must have an adequate bandwidth, usually 10 MHz or more, and an input impedance of around 10 MΩ in parallel with 30 pF (often with probes that provide a much lower capacitance). Oscilloscope testing is most useful when the shape of the correct waveform at each point is known, and manufacturers will often provide photographs of oscillograms to illustrate the waveshapes, particularly for TV equipment.

Where such waveshapes are not known, the circuit needs to be more thoroughly understood and in some cases, the CRO is not an adequate guide. Distortion in audio equipment, for example, needs to reach intolerable levels to become obvious on an oscilloscope display of a sine wave, and a specialised distortion meter (which filters and detects harmonics) must be used. For many other applications, however, and particularly in pulse timing measurements, the CRO is a valuable method of diagnosis for linear circuits, and a good general checking method for digital circuitry. More specialised methods are usually necessary for microprocessor circuitry, however.

Practical work on microprocessor equipment

In some respects, servicing microprocessor circuitry is simpler than work on analogue circuits of comparable size. All digital signals are at one of two levels, and there is no problem in identifying minor changes of waveshape which so often cause trouble in audio circuits. In addition, the specifications that have to be met by a microprocessor circuit can be expressed in less ambiguous terms than those that have to be used for analogue circuits. There is no need, for example, to worry about harmonic distortion or intermodulation. That said, microprocessor circuits bring their own particular headaches, the worst of which is the relative timing of voltage changes.

Digital signals are generally changing with each clock pulse, and the problem is compounded by the fast clock rates that have to be used for many types of microprocessors. An example is to find a coincidence of two pulses with a 12 MHz clock pulse when the coinciding pulses happen only when a particular action is taking place. This action may be completely masked by many others on the same lines, and it is in this respect that the conventional oscilloscope is least useful. Oscilloscopes as used in analogue circuits are intended to display repetitive waveforms, and are not particularly useful for displaying a waveform which once in 300 cycles shows a different pattern. The conventional oscilloscope is useful for checking pulse rise and fall times, and for a few other measurements, but for anything that involves bus actions a good storage oscilloscope is needed. In addition, some more specialised equipment will be necessary if anything other than fairly simple work is to be contemplated. Most of this work is likely to be on machine-control circuits and the larger types of computers. Small computers do not offer sufficient profit margin in repair work to justify much diagnostic equipment. After all, there is not much point in carrying out a £200 repair on a machine which is being discounted in the shops to £150. Spare parts for the smaller home computers are also a source of worry, because many are custom made, and may not be available by the time that the machines start to fail.

Failures in microprocessor circuits are either of components, open or short circuit, or in i.c.s. For ordinary logic circuits, using TTL or CMOS chips, one very useful diagnostic method relies on slow clocking. The state of buses and other logic lines can be examined using LEDs, and with a 1 second or slower clock rate, the sequence of signals on the lines can be examined from any starting state. This approach is not usually available for microprocessor circuits. A few microprocessors of CMOS construction, like the Intel CHMOS 80C86, can be operated at very slow clock rates, down to DC. This is a very useful feature, though it does not necessarily help unless you can get the buses into the state at which the problem reveals itself. As in any other branch of servicing, your work is made very much easier if you have some idea of where the fault may lie. For example, if the user states that short programs run but long ones crash, this is a good pointer to something wrong in the higher-order address lines, such as an open-circuit contact on an i.c. holder for a memory chip. Slow clocking is not necessarily helpful for such problems, because a large number of clock cycles may be needed to reach the problem address by hardware.

For many aspects of fault finding, the use of a diagnostic program is very helpful. Such programs are usually available for computer servicing, and will help to pinpoint the area of the problem. A test program can find problems in the CPU, the RAM and the ROM, and can also point to faults in the ports or in decoder chips. Some machines, such as the IBM PC, run a short diagnostic program each time the machine is switched on, and more extensive programs can be obtained to order.

Diagnostic programs are not always available for machine-control systems, because so many systems are custom designed. If it is likely that one particular system design will turn up several times, then a simple diagnostic program should be written, enlisting a software specialist if necessary. One very simple method is to use code that consists entirely of NOP (No Operation) instructions, since this will allow the action of the address lines to be studied as they cycle through all the available addresses.

For small computers, the use of a good diagnostic program may be all that is normally needed to locate a fault. This is particularly true when the machine is one that has a reasonably long service record, with well-documented problems and their solutions available for a large sample of machines. Servicing the old BBC Micro, for example, is made much easier by the service history which has been built up by local education authorities on this machine, and servicing of such a standardised product as the older versions of IBM clone machines is always comparatively simple. By contrast, servicing of a comparatively new model may be virtually impossible through lack of information, and manufacturers can be quite remarkably uncooperative both in the provision of information and of spares. This is very often because they are working on the next model and have lost interest in last month's wonder. For machine-control circuits, easy availability of either data or spares cannot be relied on, and servicing may have to be undertaken with little more than a circuit diagram, and the datasheets for the chips that are used. The compensation here is that the chips are more likely to be standard types, with no custom-built specials.

The main problem in this respect is that servicing of microprocessor circuits cannot ever be a purely hardware operation. Every action of a microprocessor system is software controlled, and in the course of fault diagnosis, a program must be running. For a computer system, this is easily arranged, and a diagnostic program can be used, but for a machine-control system this is by no means simple. A machine-control system, for example, may have to be serviced *in situ*, simply because it would be too difficult to provide simulated inputs and outputs. In such circumstances, dummy loads may have to be provided for some outputs to avoid unwanted mechanical actions. Against this should be laid the point that inputs and outputs are more easily detectable, and likely to be present for longer times, making diagnosis rather easier unless the system is very complicated. A logic diagram, showing the conditions that have to be fulfilled for each action, is very useful in this type of work. Once again, if the system is one that would have to be serviced frequently, then a 'dummy driver' which will provide simulated inputs, can be a very useful service tool.

An important point to note in this respect is the number of ways that the word **monitor** is used. The conventional meaning of the noun **monitor** is the VDU or screen display unit, and the verb **to monitor** means to observe and check a quantity. In software work, however, a **monitor** (more correctly a monitor program) is a form of diagnostic program which will show information such as the contents of CPU registers and system memory along with contents of specified disk sections, allowing such contents to be changed, and the running of programs step by step. Such a program is very useful for tracing software faults.

Instruments for digital servicing work

Of the more specialised instruments which are available for working with microprocessor circuits, logic probes, pulsers, monitors and logic analysers are by far the most widely used. A logic probe is a device which uses a small conducting probe to investigate the logic state of a single line. The state of the line is indicated by LEDs, which will indicate high, low, or pulsing signals on the line. The probe is of very high impedance, so that the loading on the line is negligible.

These probes are not costly, and are extremely useful for a wide range of work on faults of the simpler type. They will not, obviously, detect problems of mistiming in bus lines, but such faults are rare if a circuit has been correctly designed in the first place. Most straightforward circuit problems, which are mainly chip faults or open or short circuits, can be discovered by the intelligent use of a logic probe, and since the probe is a pocket-sized instrument it is particularly useful for on-site servicing. Obviously, the probe, like the voltmeter used in an analogue circuit, has to be used along with some knowledge of the circuit. You cannot expect to gain much from simply probing each line of an unknown circuit. For a circuit about which little is known, though, some probing on the pins of the microprocessor can be very revealing. Since there are a limited number of microprocessor types, it is possible to carry around a set of pinouts for all the microprocessors that will be encountered.

Starting with the most obvious point, the probe will reveal whether a clock pulse is present or not. Quite a surprising number of defective systems go down with this simple fault, which is more common if the clock circuits are external. Other very obvious points to look for are a permanent activating voltage on a HALT line, or a permanent interrupt voltage, caused by short circuits. For an intermittently functioning or partly functioning circuit, failure to find pulsing voltages on the higher address lines or on data lines may point to microprocessor or circuit-board faults. For computers, the description of the fault condition along with knowledge of the service history may be enough to lead to a test of the line that is at fault. The considerable advantage of using logic probes is that they do not interfere with the circuit, are very unlikely to cause problems by their use, and are simple to use. Some 90% of microprocessor system faults are detectable by the use of logic probes, and they should always be the first hardware diagnostic tool that is brought into action against a troublesome circuit. A variant on the logic probe is the logic clip which as the name suggests clips on to the line and avoids the need to hold the unit in place. Logic clipping can also be used on chips, using a clip unit which fastens over the pins of the chip and allows probes to be connected with no risk of shorting one pin to another.

Logic pulsers are the companion device to the logic probe. Since the whole of a microprocessor system is software operated, some lines may never be active unless a suitable section of program happens to be running. In machine-control circuits particularly, this piece of program may not run during any test, and some way will have to be found to test the lines for correct action. A digital pulser, as the name indicates, will pulse a line briefly, almost irrespective of the loading effect of the chips attached to the line. The injected pulse can then be detected by the logic probe. This method is particularly useful in tracing the path of a pulse through several gate and flip-flop stages, but not if several conditions have to be fulfilled at any one time. The digital pulser is a more specialised device than the logic probe, and it has to be used with more care. It can, however, be very useful, particularly where a diagnostic program is not available, or for testing actions that cannot readily be simulated.

The logic comparator (or logic monitor) is an extension of the logic probe to cover more than one line. The usual logic monitor action extends to 16 lines, with an LED indicating the state of each line. The LED is lit for logic high, and unlit for logic zero or the floating state. For a pulsing line, the brightness of each LED is proportional to the duty cycle of the pulses on the line. Most logic monitors have variable threshold voltage control, so that the voltage of transition between logic levels can be selected to eliminate possible spurious levels. Connection to the circuit is made through ribbon cable, terminating usually in a clamp which can be placed over an i.c. For microprocessor circuits, the usual clamp is a 40-pin type, and the ends of the ribbon cable must be attached to the set of bus lines for that particular microprocessor. Pre-wired clamps are often available for popular microprocessor types. For microprocessor circuits, however, the use of a 40-point monitor is much better, since only a knowledge of the microprocessor pinout will then be needed. Since port chips are generally in a 40-pin package also, this allows tests on ports, which are often a fruitful source of microprocessor system troubles.

Logic analysers

The logic analyser is an instrument which is designed for much more detailed and searching tests on digital circuits generally and microprocessor circuits in particular. As we have noted, the conventional oscilloscope is of limited use in microprocessor circuits because of the constantly changing signals on the buses as the microprocessor steps through its program. Storage oscilloscopes allow relative timing of transitions to be examined for a limited number of channels, but suitable triggering is seldom available. Logic probes and monitors are useful for checking logic conditions, but are not useful if the fault is one that concerns the timing of signals on different lines. The logic analyser is intended to overcome these problems by allowing a time sample of voltages on many lines to be obtained, stored, and examined at leisure.

Most logic analysers permit two types of display. One is the 'timing diagram' display, in which the various logic levels for each line are displayed in sequence, running from left to right on the output screen of the analyser. A more graphical form of this display can be obtained by connecting a conventional oscilloscope, in which case the pattern will resemble that which would be obtained from a 16-channel storage oscilloscope. The synchronisation may be from the clock of the microprocessor system, or at independent (and higher) clock rates which are more suited for displaying how signal levels change with time. The other form of display is word display. This uses a reading of all the sampled signals at each clock edge, and displays the results as a 'word', rather than as a waveform. If the display is in binary, then the word will show directly the 0 and 1 levels on the various lines. For many purposes, display of the status word in other forms, such as hex, octal, denary or ASCII may be appropriate. This display, which gives rise to a list of words as the system operates, is often better suited for work on a system that uses buses, such as any microprocessor system. The triggering of either type of display may be at a single voltage transition, like the triggering of an oscilloscope, or it may be gated by some preset group of signals, such as an address. This allows for detecting problems that arise when one particular address is used, or one particular instruction executed.

A current tracer is a way of measuring current in a PCB track without the need to make a gap in the track and attach a meter. The tracer works either by sensing the magnetic field around the track or by measuring the voltage drop along a piece of track, and measurements are accurate only if the PCB track conforms to the standard 306 g/m² density specification, and with the tracer set for the correct width of track. Currents of 10 mA or less can be measured, and this is a very useful way of finding cracked tracks, dry joints, shorted tracks and open circuit at plated-through connections.

Finally, a signature-checker allows for faults in a ROM to be checked. When a prototype system has been constructed and proved to function correctly it can be made to execute a sequence of instructions in a repetitive manner. By monitoring each node, data about the correct logical activity can be accumulated. By adding such data to the circuit or system diagram, a 'signature' for each node, under working conditions, is provided. In servicing, a faulty component can then be identified as a device that produces an error output from correct input signatures.

For computer ROMs, the definition is slightly different. When the ROM is perfect, adding all of the bytes will produce a number which is the 'signature' for that ROM, and repeating the addition should always produce the same signature if the ROM is perfect. Any deviation will indicate a fault which can be cured only by replacement of the ROM. The signature is usually obtained by a more devious method than has been suggested, keeping the size of the sum constant and omitting any overflow bits, but the principle is the same.

Appendix I

Standard metric wire table

Diameter (mm)	Resistance (ohms per metre)	Current rating (mA)
0.025	35.1	2.3
0.032	21.4	3.7
0.040	13.7	5.8
0.050	8.8	9.1
0.063	5.5	14.5
0.080	3.4	23.4
0.100	2.2	36.5
0.125	1.4	57.1
0.140	1.1	71.6
0.160	0.86	93.5
0.180	0.68	118.3
0.200	0.55	146.1
0.250	0.35	228.3
0.280	0.28	286.3
0.315	0.22	362.4
0.400	0.14	584.3
0.450	0.11	739.5
0.500	0.088	913.0
0.56	0.070	1.14 (amperes)
0.63	0.055	1.45
0 71	0.043	1.84
0.75	0.039	2.05
0.80	0.034	2.34
0.085	0.030	2.64
0.90	0.027	2.96
0 95	0.024	3.30
1.00	0.022	3.65

The values of resistance per metre and of current rating have been rounded off. Only the smaller gauges are tabulated, representing the range of wire gauges which might be used in constructing r.f. and a.f. transformers.

Appendix II

Bibliography

Some of the most useful reference books in electronics are either out of print or difficult to obtain. They have been included in the list below because they can often be found in libraries, or in second-hand shops.

Components: Understanding Electronic Components (Sinclair) Fountain Press, 1972.

Formulae and tables: Reference Data for Radio Engineers (ITT).

Audio & Radio: Radio Designers's Handbook (Langford-Smith) Iliffe, 4th edn, 1967. A wealth of data, though often on valve circuits. Despite the age of the book, now out of print, it is still the most useful source book for audio work.

Radio Amateurs Handbook (ARRL). A mine of information of transmitting and receiving circuits. An excellent British counterpart is available, but the US publication contains more varied circuits, because the US amateur is not so restricted in his operations.

GE Transistor Manual (General Electric of USA). Even the early editions are extremely useful.

Oscilloscopes: The Oscilloscope in Use (Sinclair) Argus Books, 1976.

Manufacturer's Databooks by Texas, RCA, SGS—ATES, Motorola, National Semiconductor and Mullard contain detailed information on semiconductors, with many applications circuits.

Microprocessors: The textbooks by Dr. Adam Osborne are by far the most useful for anyone who already has some knowledge of microprocessors. For beginners, the last chapter of *Beginner's Guide to Digital Electronics* is a useful introduction.

A useful book on home computers and programming is *Microprocessors for Hobbyists* (Coles) Newnes Technical Books, 1979.

Other, up-to-date, titles from Newnes Technical Books include the following:

Newnes Radio and Electronics Engineer's Pocket Book (16th Edition), 1986, by Keith Brindley.

The Practical Electronics Microprocessor Handbook, 1986, by Ray Coles.

16-Bit Microprocessor Handbook, 1986, by Trevor Raven.

Oscilloscopes: how to use them, how they work (2nd Edition), 1986, by Ian Hickman.

Op-Amps: their principles and applications (2nd Edition), 1986, by J. Brian Dance.

The Art of Micro Design, 1984, by A.A. Berk.

Practical Design of Digital Circuits, 1983, by Ian Kampel.

In addition there are *Beginner's Guides* on the following subjects: *Amateur Radio, Hi-Fi, Computers, Digital Electronics, Electronics, Microcomputing, Microprocessors, Radio, Television, Video,* and *Video-cassette Recorders.*

For further information on Newnes Technical titles write to the publishers for a catalogue.

Appendix III

The Hex Scale

Hexadecimal means scale of 16, and the reason it is used so extensively is that it is naturally suited to representing binary bytes. Four bits, half of a byte, will represent numbers which lie in the range 0 to 15 in our ordinary denary number scale. This is the range of one hex digit (*Figure AIII.1*). Since we do not have symbols for digits higher than 9, we have to use the letters A, B, C, D, E, and F to supplement the digits 0 to 9 in the hex scale. The advantage is that a byte can be represented by a two-digit number, and a complete address by a four-digit number. Converting between binary and hex is much simpler than converting between binary and denary. The number that we write as 10 (ten) in denary is written as 0A in hex, 11 as 0B, 12 as 0C and so on up to 15, which is 0F. The zero does not have to be written, but programmers get into the habit of writing a data byte with two digits and an address with four even if fewer digits are needed. In this brief description, only single-byte numbers will be used; the principles are the same for words and Dwords.

The number that follows 0F is 10, 16 in denary, and the scale then repeats to 1F, 31, which is followed by 20. The maximum size of byte, 255 in denary, is FF in hex. When we write hex numbers, it is usual to mark them in some way so that you do not confuse them with denary numbers, There is not much chance of confusing a number like 3E with a denary number, but a number like 26 might be hex or denary. The convention that is followed by many programmers is to use a capital H to mark a hex number, with the H-sign placed after the number. For example, the number 47H means hex 47, but plain 47 would mean denary forty-seven.

Hex	Denary	Hex	Denary
0	0	B	11
1	1	C	12
2	2	D	13
3	3	E	14
4	4	F	15
5	5	then	
6	6	10	16
7	7	11	17
8	8	to	
9	9	20	32
A	10	21	33
		etc.	

Figure AIII.1

Hex	Binary	Hex	Binary
0	0000	8	1000
1	0001	9	1001
2	0010	A	1010
3	0011	B	1011
4	0100	C	1100
5	0101	D	1101
6	0110	E	1110
7	0111	F	1111

Figure AIII.2

Another method is to use the hashmark before the number, so that #47 would mean the same as 47H.

The great value of hex code is how closely it corresponds to binary code. If you look at the hex–binary table of *Figure AIII.2*, you can see that #9 is 1001 in binary and #F is 1111. The hex number #9F is therefore just 10011111 in binary – you simply write down the binary digits that correspond to the hex digits. Taking another example, the hex byte #B8 is 10111000, because #B is 1011 and #8 is 1000. Conversion in the opposite direction is just as easy – you group the binary digits in fours, starting at the least significant (right-hand) side of the number, and then convert each group into its corresponding hex digit. *Figure AIII.3* shows examples of the conversion in each direction.

Negative numbers

There is no negative sign in hex arithmetic, nor in binary either. The conversion of a number to its negative form is done by a method called complementing, and *Figure AIII.4* shows how this is done. At first sight,

Conversion: Hex to Binary

Example: 2CH . . . . 2H is 0010 binary
                                CH is 1100 binary
so that 2CH is 00101100 binary
Example: 4A7FH . . . . 4H is 0100 binary
                                AH is 1010 binary
                                7H is 0111 binary
                                FH is 1111 binary
so that 4A7FH is 0100101001111111 binary

Conversion: Binary to hex

Example: 01101011 . . . . 0110 is 6H
                                1011 is BH
so that 01101011 is 6BH
Example: 1011010010010 – note that this is not a complete number of bytes.
Group into fours, starting with the least significant byte.
                                0010 is 2H
                                1001 is 9H
                                0110 is 6H
                                1 is 1H
so that the hex equivalent (reading bottom to top this time) is 1692H

Figure AIII.3

Binary number . . . . 00110110 (denary 36)
inverted (complemented) 11001001
add 1 . . . . 11001010 (denary 202 or −36)

Denary number −5
In binary +5 is 00000101 (in 8-bit binary)
inverted 11111010
add 1 . . . . 11111011 which is −5 (or 251)

Note that the value of the denary equivalent depends on whether the binary
number is taken as being signed or unsigned.

Figure AIII.4

and very often at second, third, and fourth, it looks entirely crazy. When
you are dealing with a single-byte number, for example, the denary form
of the number −1 is 255. You are using a large positive number to
represent a small negative one! It begins to make more sense when you
look at the numbers written in binary. The 8-bit numbers that can be
regarded as negative all start with a 1 and the positive numbers all start
with a 0. The CPU can find out which is which just by testing the
left-hand bit, the most significant bit, often called the **sign bit**. It is a
simple method, which the machine can use efficiently, but it does have
disadvantages for mere humans. One of these disadvantages is that the

digits of a negative number are not the same as those of a positive number. For example, in denary −40 uses the same digits as +40; in hex, −40 becomes D8H and +40 becomes 28H. The denary number −85 becomes ABH and +85 becomes 55H. It is not at all obvious that one is the negative form of the other.

The second disadvantage is that humans cannot distinguish between a single-byte number which is intended to be negative and one which is just a byte whose value is greater than 127. For example, does 9FH mean +159 or does it mean −97? The short and remarkable answer is that the human operator does not have to worry. The microprocessor will use the number correctly no matter how we happen to think of it. The snag is that we have to know what this correct use is in each case. Throughout this book, and in others that deal with machine code programming, you will see the words 'signed' and 'unsigned' used. A signed number is one that may be negative or positive. For a single-byte number, values of 0 to 7FH are positive, and values of 80H to FFH are negative. This corresponds to denary numbers 0 to 127 for positive values and 128 to 255 for negative. Unsigned numbers are always taken as positive.

If you find the number 9CH described as signed, then you know it is treated as a negative number (it is more than 80H). If it is described as unsigned, then it is positive, and its value is obtained simply by converting. The snag here is that when we make use of the H command of DEBUG, it will not deal with signs in single bytes. If, for example, you type:

H 2A 2B

then what you will see underneath is 0055, the sum, and FFFF, which is the difference. You get FFFF rather than FF, because the H command works with hex numbers of four digits, word size. It is not a real problem, because all you have to do is to ignore the first two digits when you are working with single bytes.

Index